Advances in Pre- and Post-Additive Manufacturing Processes

This book provides knowledge about the process of creating and designing products based on an Industry 4.0 setting. The fundamentals of Additive Manufacturing, its many technologies, the process parameters, advantages, limitations, and recent developments are discussed. In addition, the most recent post-additive manufacturing process advancements, surface quality defects, and challenges are the primary topics that will be investigated in the book.

Advances in Pre- and Post-Additive Manufacturing Processes: Innovations and Applications provides scientific and technological insights into the physical fundamentals of the machining and finishing processes in macro, micro, and nanoscales. It explores in a systematic way both conventional and unconventional material-shaping processes with various modes of hybridization concerning theory modelling and industrial potential. It focuses on the applications of Additive Manufacturing that are linked to pre-stage and post-stage processes and encompasses a broad spectrum of macro, micro, and nano-processes that are utilized in manufacturing activities. The book goes on to cover a wide range of reliable and economical fabrication of metallic parts with complicated geometries which are of considerable interest to the aerospace, medical, automotive, tooling, and consumer products industries.

This reference title encapsulates the current trends of today's material development and machining techniques for advanced composite materials, making it a one-stop resource for academic researchers and industrial firms while they are formulating strategic development strategies. It also serves as a reference book for students at all levels of education, from undergraduates to doctoral candidates.

Innovations in Smart Manufacturing for Long-Term Development and Growth
Series Editors: Atul Babbar, Gursel Alici, Yu Dong, Ankit Sharma

Fabrication Techniques and Machining Methods of Advanced Composite Materials

Vikas Dhawan, Atul Babbar, Inderdeep Singh and Jonathan M. Weaver

Advances in Pre- and Post-Additive Manufacturing Processes

Innovations and Applications

Naveen Mani Tripathi and Ankit Sharma

For more information about this series, please visit: www.routledge.com/Innovations-in-Smart
-Manufacturing-for-Long-Term-Development-and-Growth/book-series/CRCISMLDG

Advances in Pre- and Post-Additive Manufacturing Processes

Innovations and Applications

Edited by
Naveen Mani Tripathi
and Ankit Sharma

CRC Press
Taylor & Francis Group
Boca Raton London New York

CRC Press is an imprint of the
Taylor & Francis Group, an **informa** business

Designed cover image: Shutterstock - lucadp

MATLAB® and Simulink® are trademarks of The MathWorks, Inc. and are used with permission. The MathWorks does not warrant the accuracy of the text or exercises in this book. This book's use or discussion of MATLAB® or Simulink® software or related products does not constitute endorsement or sponsorship by The MathWorks of a particular pedagogical approach or particular use of the MATLAB® and Simulink® software.

First edition published 2024
by CRC Press
2385 NW Executive Center Drive, Suite 320, Boca Raton FL 33431

and by CRC Press
4 Park Square, Milton Park, Abingdon, Oxon, OX14 4RN

CRC Press is an imprint of Taylor & Francis Group, LLC

© 2024 selection and editorial matter, Naveen Mani Tripathi and Ankit Sharma; individual chapters, the contributors

ISBN: 978-1-032-54987-3 (hbk)
ISBN: 978-1-032-55066-4 (pbk)
ISBN: 978-1-003-42886-2 (ebk)

DOI: 10.1201/9781003428862

Typeset in Times
by Deanta Global Publishing Services, Chennai, India

Dr. Naveen Mani Tripathi dedicates this book to Smt. Vijay Lakshmi Mani Tripathi (mother), Shree Awalesh Mani Tripathi (father), Smt. Meenakshi Mani Tripathi (wife), and Nameesh Mani Tripathi (lovely son).

Dr. Ankit Sharma dedicates this book to Dr. Harshita Sharma (wife) and Mishka Sharma (daughter).

Contents

List of Figures and Tables

FIGURES

TABLES

Preface

This book covers the concepts, principles, choices for material and equipment, technologies, most recent innovations, linked to "pre-stage" and "post-stage" processes and applications of Additive Manufacturing (AM), also known as 3D printing. The purpose of this book is to provide knowledge about the process of creating and designing products based on an Industry 4.0 setting. The fundamentals of AM, its technologies, the process parameters, case studies, various modes of hybridization in relation to theory, modelling, and industrial potential and recent developments are also discussed in this book. In addition, the most recent post-additive manufacturing process advancements, surface quality defects, and challenges are the primary subjects that are investigated. This book encapsulates the current trends of today's material development and machining techniques for advanced composite materials, making it a one-stop resource for academics and manufacturing experts, engineers working in related industries, academic researchers, students, and so on. It will also serve them as a primary source of information while they are formulating strategic development strategies. It includes the research input of AM trends and technologies from more than nine countries of different continents, which justifies the diversities comprehended in this book. A wide range of reliable and economical fabrication of parts with complicated geometries which is of considerable interest for aerospace, medical, automotive, tooling, and consumer products industries is covered.

The book acts as a first-hand reference for commercial organizations mimicking modern material used for making additive manufactured parts and machining and finishing processes-based applications. By capturing the current trends of today's manufacturing practices, this book becomes a one-stop resource for scholars and manufacturing professionals, engineers in related disciplines, academic researchers, etc.

As a result, we have a great deal of faith that this contribution will be of use to the entirety of the readers in a variety of ways.

About the Editors

Dr Naveen Mani Tripathi is currently working as Assistant Professor in the Department of Mechanical Engineering at Assam Energy Institute, Sivasagar, Assam (A Centre of Rajiv Gandhi Institute of Petroleum Technology, Jais, Amethi, Uttar Pradesh), under the Ministry of Petroleum and Natural Gas, Government of India. He completed his Doctoral Research from the Ben-Gurion University of the Negev, Beer-Sheva, Israel. After his PhD, he worked as Particle Scientist in the GranuTools sprl, Belgium. His Post-Doctoral research is from the Chair of Process System Engineering at Technical University of Munich, Germany. He is Master's from Thapar Institute of Engineering & Technology, Patiala, Punjab, India. He has been awarded with many fellowships, such as Graduate Aptitude Taste of Engineering (GATE), India; Graduate Record Examinations (GRE) for International Studies and Fellowship, Krietman Fellowship of Israel; Research Mobility Fund through Technical University of Munich (TUM), Germany; TUM Foundation Fellowship (TUFF), Germany; and Alexander von Humboldt Fellowship, Germany. His area of research is Thermo-Fluid, Material Characterization, Additive Manufacturing, Multiphase Flow and Pipeline Design for Energy and Process Industries. He holds many national and international sponsored projects. He has three granted patents, one published patent and more than 40 journals/conference/chapter publications. He is voluntarily working as editorial board member and review editor for more than 20 journals. He is life member of many prominent institutions such as Indian Institute of Chemical Engineers (IIChE), Indian Society of Mechanical Engineers (ISME), Natural Gas Society etc.

Dr Ankit Sharma holds the position of Associate Director (Research) at the CURIN Department of Chitkara University, Punjab, India. He also heads the Chikara University Publications/publisher for international peer-reviewed Chitkara journals. He received his doctorate in Mechanical Engineering from the Thapar Institute of Engineering and Technology, Punjab, India. He is a passionate researcher with diversified research interests in manufacturing, medical devices, modern machining processes, additive manufacturing, tissue engineering, developing automated machines for health industries, and not limited to that. To date, he has published more than 55 scientific articles in various peer-reviewed, top-notch journals, conferences, and books in manufacturing and materials.

Dr. Sharma is also organizing an international conference and serving as the General chair and convenor of the International Conference on Emerging Materials, Smart Manufacturing, & 3D Printing, Computational Intelligence (ICEMSMCI 2023) with publishing partners Springer, Taylor & Francis, Inder Science, AIP Proceedings, and Bentham Science. He has also been the organizing committee member of the International Conference on the Advances in Materials and Manufacturing Technology-2022. He has chaired various international conferences in the domain of advanced manufacturing. He has delivered several seminars and keynote talks on international (USA, China, India) platforms and awarded with best research paper awards

Dr. Sharma is working on research commercialization and filed more than 30 patents. He also granted 4 international patents in his account along with 18+ granted patents. Dr. Sharma edited/authored 3 books, serving as a series editor of Taylor & Francis and ASME books. He is also serving as a guest editor and Managing editor of international peer-reviewed journals and books.

List of Contributors

K Annamalai
Department of Mechanical Engineering
VIT University
Chennai, Tamil Nadu, India

Samim Ali
ICMR – National Institute of Cancer
Prevention and Research
Noida, Uttar Pradesh, India

Atul Babbar
Department of Mechanical Engineering
SGT University
Gurugram, Haryana, India

Desai Dhaval Jaydev Kumar
Worley ECR (Energy | Chemicals |
    Resources)
Texas, USA

Ismail Fidan
Department of Manufacturing and
    Engineering Technology
Tennessee Technological University
Cookeville, USA

V M Gobinath
Department of Mechanical
    Engineering
Rajalakshmi Institute of Technology
Tamil Nadu, India

Yash Gopal Mittal
Department of Mechanical
    Engineering
Indian Institute of Technology (IIT)
Bombay
Mumbai, Maharashtra, India

Gopal Gote
Department of Mechanical
    Engineering
Indian Institute of Technology (IIT)
    Bombay
Mumbai, Maharashtra, India

Barun Haldar
Department of Mechanical and
    Industrial Engineering
College of Engineering
Imam Mohammad Ibn Saud Islamic
    University (IMSIU)
Kingdom of Saudi Arabia

Mufaddal Huzefa Shakir
Department of Automobile Engineering
Chandigarh University
Mohali, Punjab, India

Pushkar Kamble
ICB-CO2M Laboratory
University of Technology Belfort-
    Montbeliard (Sevenans)
Belfort, Sevenans, France

A Kathirvel
Department of Computer Science and
    Engineering
Panimalar Engineering College
Chennai, Tamil Nadu, India

K P Karunakaran
Department of Mechanical
    Engineering
Indian Institute of Technology (IIT)
    Bombay
Mumbai, Maharashtra, India

Ravi Kumar
Department of Mechanical Engineering
BTKIT Dwarahat
Almora, Uttarakhand, India

Avinash Kumar Mehta
Department of Mechanical Engineering
Indian Institute of Technology (IIT)
    Bombay
Mumbai, Maharashtra, India

Vijay Kumar Sharma
Department of Physics
Shyam Lal College
University of Delhi
Shahdara, Delhi, India

Akant Kumar Singh
Department of Mechanical Engineering
Chandigarh University
Mohali, Punjab, India

Paola Leo
Department of Innovation and
    Engineering
University of Salento
Lecce, Italy

Naveen Mani Tripathi
Assam Energy Institute, Sivasagar,
Assam (A Centre of Rajiv Gandhi
Institute of Petroleum Technology),
    MoPNG
Sivasagar, Assam, India

Vivek Mani Tripathi
School of Nano Science (SNS) Central
University of Gujarat
Gandhinagar, Gujarat, India

Natarajan N
Department of Mechanical Engineering
Muthayammal Engineering College
Rasipuram, Tamil Nadu, India

Satishkumar P
Department of Mechanical Engineering
Rathinam Technical Campus
Coimbatore, Tamil Nadu, India

Vikas Pandey
Department of Energy Sciences
National Institute of Natural Sciences
Okazaki, Japan

Yogesh Patil
Department of Operations and Supply
    Chain Management
Indian Institute of Management (IIM)
Mumbai, Maharashtra, India

Vikas Raghuvanshi
Madurai Kamaraj University
Palkalai Nagar
Madurai, Tamil Nadu, India

Rahutosh Ranjan
Mahatma Gandhi Central University
Motihari, Bihar, India

Gilda Renna
Department of Innovation and
    Engineering
University of Salento

Mohd Rizwan Jafar
Department of Mechanical Engineering
Delhi Technological University
Delhi, India

Seenivasan S
Department of Mechanical Engineering
Rathinam Technical Campus
Coimbatore, Tamil Nadu, India

Daniel Schiochet Nasato
Chair of Process System Engineering
Technical University of Munich
Freising, Germany

Ankit Sharma
Chitkara University Institute of
    Engineering & Technology
Chitkara University
Punjab, India

Hariom Sharma
National Institute of Health
Bethesda, MD, USA

Siddhartha
Department of Mechanical
    Engineering
National Institute of Technology
Hamirpur
Hamirpur, Himachal Pradesh, India

Dharminder Singh
Department of Mechanical
    Engineering
Glasgow Caledonian University
Glasgow, United Kingdom

Deepak Singh Chauhan
Department of Microbiology and
    Immunology
Dalhousie University
Halifax, Nova Scotia, Canada

A S K Sinha
Department of Chemical Engineering
    and Biochemical Engineering
Rajiv Gandhi Institute of Petroleum
Technology
Amethi, Uttar Pradesh, India

Neetesh Soni
Department of Innovation and
    Engineering
University of Salento
Lecce, Italy

Pratibha Tiwari
Department of Biotechnology
HIMT Group of Institution
Gr. Noida, Uttar Pradesh, India

Ashok Verma
Department of Mechanical Engineering
BTKIT Dwarahat
Almora, Uttarakhand, India

Adhishree Yadav
Centre of Bioinformatics and
    Computational Biology
CSIR – Central Drug
Research Institute
Lucknow, Uttar Pradesh, India

Kusum Yadav
Department of Biochemistry
University of Lucknow
Lucknow, Uttar Pradesh, India

Manvendra Yadav
Entrepreneurship and Management
Process International (EMPI)
New Delhi, New Delhi, India

Pramod Yadav
Amity Institute of Neuropsychology
    and Neurosciences
Amity University Uttar Pradesh
Noida, Uttar Pradesh, India

Priyanka Yadav
Department of Biochemistry
University of Lucknow
Lucknow, Uttar Pradesh, India

Sanjay Yadav
Department of Mechanical Engineering
I.T.S Engineering College
Greater Noida, Uttar Pradesh, India

Acknowledgments

We express our sincere thanks to all contributors, all those friends and colleagues who have activated our norms to take up a study like this.

The editors greatly remember the contribution and support provided by the Director, RGIPT Prof. ASK Sinha, and Vice Chancellor Dr. Archana Mantri, Chitkara University, Punjab, for providing a vibrant research and academic ecosystem.

We are highly indebted to our parents, wives, and in-laws, whose support and advice have paved the way for our study and investigations.

Words are inadequate to express our sincere thanks to our kids Nameesh and Mishka, who have created a congenial atmosphere at home for our studies and writings. They are the silent observers of our works.

The authors are indebted to their venerable friends and family for many positive interactions, some of whom are Shree Ram Chandra Pandey, Dr. Mukul Chandra Bora (Director, DUIET, Dibrugarh), Dr. Raktim Patar (Professor, JNU, New Delhi), Dr. S.S Das (Professor, Dibrugarh University), Mr. Uday Sankar Saikia, Dr. Vikas Pandey (Scientist, Japan), Dr. Hariom Sharma (Scientist, NIH, USA), and Mr. Sharda Dutt Singh.

Last but not least, the editors are highly obliged to Prof. Haim Kalman and Prof. Avi Levy (Ben-Gurion University of the Negev, Beer-Sheva, Israel), and Prof. S.S. Mallick (Thapar Institute of Engineering and Technology, Patiala, Punjab, India) whose teaching and guidance have made them capable to dive in the research area.

Dr. Naveen Mani Tripathi
Dr. Ankit Sharma

1 A Review on Additive Manufacturing in Industrial Environment

A. Kathirvel, V. M. Gobinath, and K. Annamalai

1.1 INTRODUCTION

1.1.1 OBJECTIVES

1) To offer a focused review on a novel evolving kind of hybridization which originates on MAM.
2) To address MAM's constraints of low productivity, poor surface quality, metallurgical flaws, and lack of dimensional accuracy by combining it with other production technologies.
3) To enhance flexibility in traditional manufacturing procedures; the quantity of material expense is reduced.

1.2 OVERVIEW

Hybrid Manufacturing (HM) is a combination of multiple techniques that help overcome individual limitations and that take advantage of intrinsic properties (Chu, 2014). HM has its roots in Subtractive Manufacturing (SM), but its applications and concepts have advanced through time, allowing it to combine alternative conventional production methods including assembling, shaping, and welding (Zhu et al., 2013; Lauwers et al., 2014).

The focus of this chapter is on an emerging hybridization type that has its roots in Metal Additive Manufacturing (MAM). In this technology, feedstock metal is added layer by layer, allowing the creation of metal parts with multifaceted geometries.

Previously, multiple thermal energy sources in combination with MAM were utilized in metal hybrid additive manufacturing along with cutting metal for improving the quality and productivity of built parts. Such developments occasioned in the monetization of the foremost Hybrid Manufacturing systems in the mid-2010s. However, in recent years, Metal Additive Manufacturing (MAM) has been combined with traditional metal-forming techniques to achieve enhanced material layer structures through localized plastic deformation. This hybrid approach not only results in improved stiffness and durability but is also employed in the production of

DOI: 10.1201/9781003428862-1

various components. Furthermore, the rapid growth of hybridization, which involves the integration of forming and MAM, has been driven by the incorporation of innovative concepts derived from both sheet and large metal-forming technologies. Some other investigations were also aimed at the cutting-edge developments related to the additive and subtractive manufacturing processes (Sharma et al., 2022; Sharma & Jain, 2020; Sharma et al., 2018; Sharma et al., 2019a; Sharma et al., 2019b; Sharma et al., 2019c; Sharma et al., 2021a; Sharma et al., 2021b; Sharma et al., 2020; Sharma et al., 2022; Kalman et al., 2017; Lumay et al., 2019; Tripathi & Mallick, 2017; Neveu et al., 2020a; Neveu et al., 2020b; Ratsimba et al., 2021; Tripathi et al., 2020; Tripathi et al., 2019; Tripathi et al., 2016; Tripathi and Mallick, 2014).

1.3   METAL ADDITIVE MANUFACTURING

Developments in additive manufacturing were first made by Hideo Kodama in the 1980s. He utilized ultraviolet lights to create solid pieces and harden the polymers (Kodama, 1981). In the late 1980s additive manufacturing was considered an emerging technology, and in the early 1990s, Charles Hull (1990) developed the stereolithography (SLA) technique and Scott Crump (1991) developed fused accumulated material. Both of these processes were utilized in the fabrication of 3D parts by thin polymer layering in flat cross-sections aided with a concentrated thermal energy supplier.

1.4   CLASSIFICATION

Manufacturing processes have been classified into seven categories by EN SO/ ASTM 52921 (2015) (Figure 1.1). From these categories, four are currently being employed to construct metal parts. These include Powder Bed Fusion (PBF), Direct Energy Deposition (DED), Sheet Lamination (SL), and Binder Jetting (BJ). When it comes to metal applications, the remaining three are classified as indirect additive manufacturing: Material Extrusion (ME), Material Jetting (MJ), and Vat Photopolymerization (VP).

1.4.1   DIRECT MAM

There are four categories in direct MAM that are in use currently to build metal parts (Figure 1.1): Powder Bed Fusion (PBF), Binder Jetting (BJ), Direct Energy Deposition (DED), and Sheet Lamination (SL).

1.4.2   INDIRECT MAM

Indirect Metal Additive Manufacturing (MAM) encompasses three primary categories for fabricating polymer-based highly filled polymer, metal matrix, and ceramic powder components (Figure 1.2): Material Jetting (MJ), Vat Photopolymerization (VP), and Material Extrusion (ME).

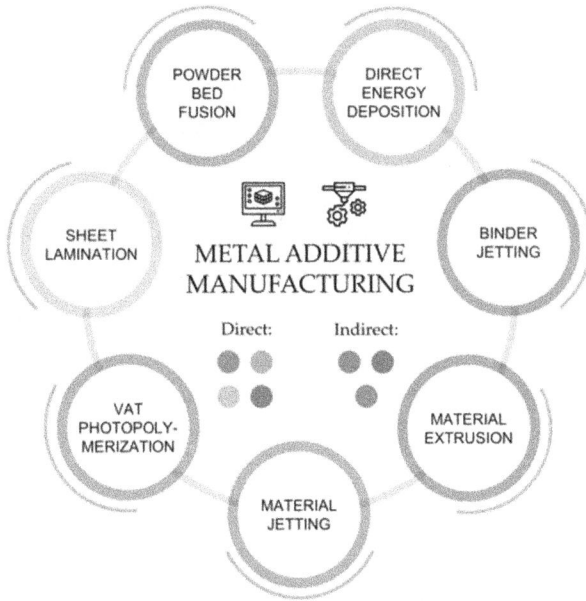

FIGURE 1.1 Additive manufacturing classification – direct and indirect appropriateness to build metal parts (adapted from the EN ISO/ASTM 52921).

1) Binder jetting
2) Sheet lamination
3) Friction surfacing
4) Powder bed fusion
5) Direct energy deposition

1.4.3 Hybrid Manufacturing

The appropriate definition for 'hybrid manufacturing' was formulated following its utilization in the late 1990s. A technology in which two or more material-removing processes are combined is termed Hybrid Manufacturing (Rajurkar et al., 1999). This definition looks imprecise to some extent because of the fact that to most subtractive routes, mixtures of two or more material elimination techniques are required; thus Kozak and Rajurkar (2000) reframed the definition according to the difference between performance features of hybrid machining processes and individual processes when they are separately performed.

Some definitions of Hybrid Manufacturing are mentioned below (Zhuet al., 2013):

(a) According to a definition, the use of multiple process mechanisms in the same processing zone is known as Hybrid Manufacturing.
(b) A general explanation of Hybrid Manufacturing is the integration of multiple manufacturing techniques into a newly merged set-up.

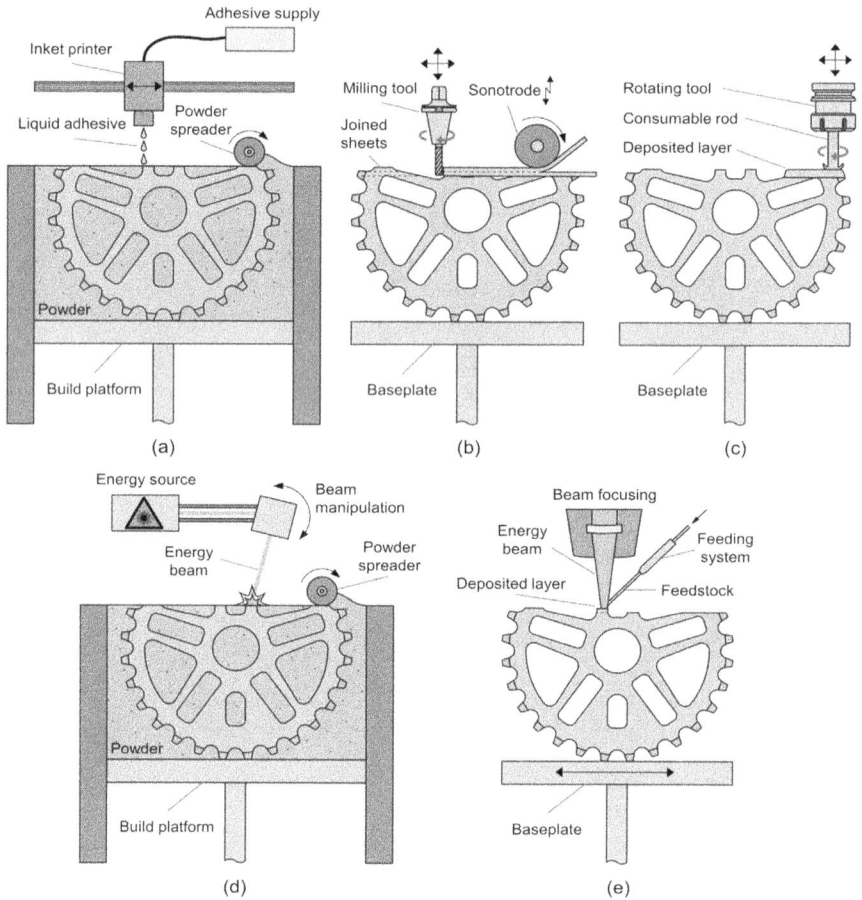

FIGURE 1.2 Diagrammatic illustration of the operational principles underlying various additive manufacturing categories.

1.5 METAL HYBRID ADDITIVE MANUFACTURING

1.5.1 METAL HAM WITH MATERIAL REMOVAL TECHNIQUES

There are two forms of hybridization of Metal Additive Manufacturing (MAM) with material removal techniques:

(a) Material elimination procedures are deployed at the post-processing stage to achieve the geometry, accuracy, surface condition, and dimensional tolerances necessary for the completed item.

(b) During a manufacturing sequence, the integration of material-eliminating processes to get components that would be very challenging and costly to generate individually either by material-eliminating operations or Additive Manufacturing.

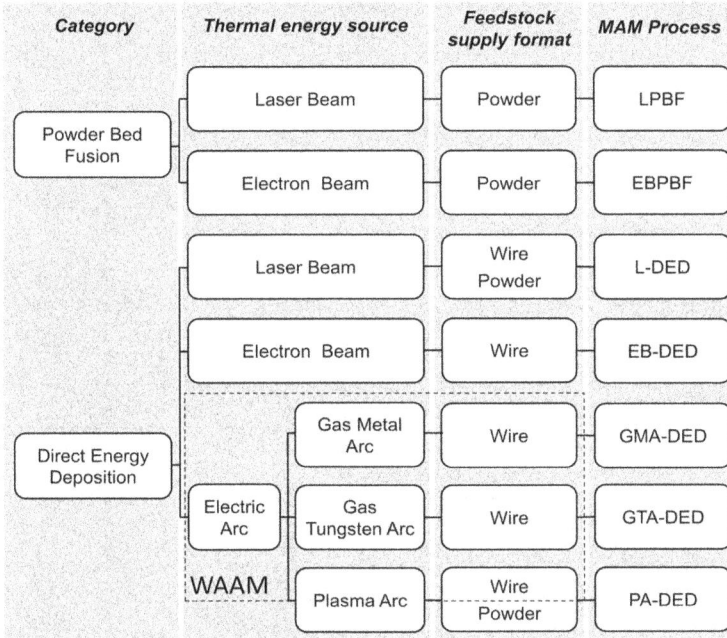

Category	Thermal energy source		Feedstock supply format	MAM Process
Powder Bed Fusion	Laser Beam		Powder	LPBF
	Electron Beam		Powder	EBPBF
Direct Energy Deposition	Laser Beam		Wire Powder	L-DED
	Electron Beam		Wire	EB-DED
	Electric Arc	Gas Metal Arc	Wire	GMA-DED
		Gas Tungsten Arc	Wire	GTA-DED
	WAAM	Plasma Arc	Wire Powder	PA-DED

FIGURE 1.3 Two categories (PBF and DED) of Metal Additive Manufacturing (MAM) have been shown here having extensive application properties in the creation of three-dimensional parts of metal.

1.5.2 METAL HAM WITH FORMING PROCESSES

Hybridizing Metal Additive Manufacturing (MAM) with forming techniques typically involves the processing of medium or large batches to produce semi-finished products, to which Additive Manufacturing incorporates additional functional elements. The conventional forming procedure routes can be extended by using this efficient methodology and can be utilized for the construction of tailor-made consumer-oriented products (Merklein, 2016).

1.6 REVIEW

Additive Manufacturing in metals that was utilized in most of the engineering material first came into use in the early 1990s along with the method called Binder Jetting (Sachs et al., 1993).

The improvements in laser technology has enabled high concentration of energy required for the processing of metal powder from when that element is in the fused position. By mid-1990s a new development came by, which was called direct metallic laser sinter (Shellabear and Nyrhilea, 2004), Along with these came the commercial utilization of gadgets for Metal Additive Manufacturing. Direct metallic laser sintering was considered as an extension of a process called selective laser sintering

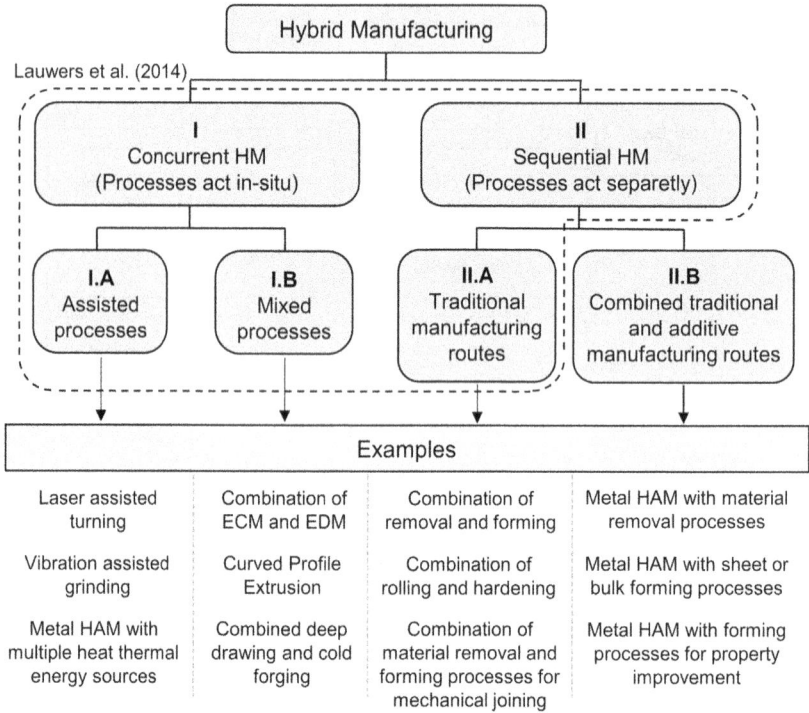

FIGURE 1.4   Hybrid manufacturing on additive manufacturing.

that was assumed to be initiated in the 1980s by a researcher Carl Deckard for the polymer material in Additive Manufacturing (Ning et al., 2005).

In late 1990s a Swedish researcher Arcam initiated the utilization of electron beam energy resources in metal-based Additive Manufacturing (Larson, 1998). Meanwhile, with the improvements in laser technology, Dickens introduced a process called 3D welding that enables them to produce exact geometrical shapes by retrofitting and the combination of robots and welding machines (Dickens et al.,1992).

The wire arc Additive Manufacturing was first developed by two persons namely, Dickens and Weisis in 1992 and 1993, respectively. The thermal energy sources were utilized in the form of electric arc and feedstock in the form of welding wire in order to produce larger parts; it's all about working procedure (Dickens et al., 1992; Prinz & Weiss 1993).

The spraying of a stagnant unfirm layer of powder positioned at a physical structure along with the adhesion by a printer with inkjet style in the hope of attaching the powder in the area together is collectively explained by a process called Binder Jetting (Sachs et al., 1993).

The rising dense property during the sintering is achieved by the contraction and failure of precision in dimension, which is analyzed as the major disadvantage of utilization of Binder Jetting (Ziaee & Crane, 2019).

The sheet joining on the top of another surface can be done by the utilization of various processes like friction stir, ultra-sonic welding or adhesive bonding. (Derazkola et al., 2020). AFS utilizes non-expandable equipment in order to produce heat and execute the integration of plastic distortion of metallic powder which oscillates via the tube with deposition on the plate at the base. The first most developed process for the coating of a surface was known as friction stir (FS) (Palanivel & Mishra, 2017).

The slicing up of a structure into various layers, followed by the summation of specific particles of grain structure on a single layer by a specified platform for a clear visualization of thermal energy resources, is termed as the metal components fabrication by Powder Bed Fusion (Bhavar et al., 2014).

The direct feeding or deposition of wire or powder by a nozzle on the top of already built components is led by the Direct Energy Deposition method, in which the thermal energy focusing melts the wire or powder to carry out the process (Saboori et al., 2017).

In comparison with Sheet Lamination and Binder Jetting, which are considered to be less commercially utilized processes, the alternative methods that are widely acceptable and applicable on a commercial scale for producing 3D metal parts are Direct Energy Deposition and Powder Bed Fusion methods (DebRoy at el., 2018).

Stereolithography (laser-based operations) and the DLP which is executed by producing a singular layer of photopolymers can be 2D shapes or patterns when integration of repeated coating is initiated (Appuhamillage et al., 2019).

For high-precision production, two photon planography can be used, where UV-evoked polymerization happens entirely in the interfered area in two beams of laser (Oran et al., 2018).

Material jetting produces components by deposition of drops of liquefied photopolymer resin stratum on the sheet by printer head of inkjet style, which are attached together through a medium of ultraviolet rays (Yap et al., 2017).

The polymers structures are produced by the technique called the drop-on-demand for investment casting, which includes the utilization of two printer heads. The first one is used for the deposition of soluble support patterns while the other is utilized for building physical materials that support the 3D printer path designing facilitation.

To build a 3D component, Material Extrusion deposition takes place on a layer-to-layer polymers of thermoplastics by drawing the polymers via a thermally heated nozzle in uninterrupted flow or stream (Gonzalez-Gutierrez et al., 2018).

To relocate or motion a laser beam reflective mirrors are utilized, with the well-defined two-dimensional way in a closed system with atmospheric and environmental conditions maintained of nitrogen and argon, which depends upon the reactivity of metals (Pragana et al., 2020).

Laser Powder Bed Fusion (LPBF) is considered to be a valid Metal Additive Manufacturing process that depends upon the final characteristics of already-built components fabricated through a wide range of metal alloys (Figure 1.3) (Bhavar et al., 2014).

The wider span utilization of these techniques, related to the development of the tools by the fabricators' alloys, results in the deposition rate ranges from 0.1 and the rough surfaceness in the limit of 10–19 μm. This concludes the overall enhancement rate in the LPBF gadgets sales annually (Wohlers, 2017). The feedstock is heated up to 0.5–0.6 of its melting point to withstand powder dispersion produced by the electrostatic charge, and this phenomenon is known as the powder push away process (Murr et al., 2012). EBPBF made its path against the specific industrial utilization, especially when it encounters the processing of hard-to-process ingredients like niobium, titanium aluminides, nickel, cobalt etc. (Keorner, 2016).

The working principle of the electron beam deposition through direct energy sources is identical to the LDED except for the laser beam sources that are operated by the thermal energy under the restricted control of vacuum conditions (Fuchs et al., 2018).

In the processing of gas metal arc welding, the electrode is implanted with the electric arc between the electrodes which is operated by a nozzle into the welded surface with the limitation of shielding gases that would be either active or inert. It is assumed to be the most widely spread technique used in WAAM. It is also a process which is called gas metal arc with the DAD (Williams et al., 2016).

The plasma arc energy and the gas tungsten arc energy deposition processes are considered to be the branches of WAAM-based techniques in which the utilization of electric arc is generated by the non-expendable electrodes (Baufeld et al., 2010), (Martina et al., 2012). Plasma arc welding and the gas tungsten arc are the two processes assumed to be the retrieving agents of the working principle of these techniques (Wu et al., 2018).

Electromagnetic forming and extrusion are various processes of the summation of forming techniques. Hybrid Manufacturing is one of the basic techniques with a combination of processes, along with the various forms of energy sources which are utilized in the identical processing region in the meantime (Figure 1.4) (Nau et al., 2011). Synchronous and restricted interaction need the technique mechanism or the availability of energy resources or equipment to react maximum or minimum in the same processing region and with the identical time for raising the proper definition of explaining the Hybrid Manufacturing (Zhu et al., 2013).

The combined techniques don't need to depend upon nonidentical technology. Araghi (2009), merged the two technologies i.e., incremental forming and stretch forming and turned them into a hybrid system (Araghi et al., 2009). The main concept alternatively centred on machining processes which involves other manufacturing techniques and paths. However, the foundation of these processes is classified by Lauwers et al. (2014).

The use of materials that are additively manufactured through this and other manufacturing techniques that were used traditionally, to manufacture components that were assumed to be difficult to produce can be achieved by several techniques individually, so the classification needs to be modified accordingly (Lauwers et al, 2014). The idea initiated the concentration in 2004 and have the foundation of improvements in hybrid welding techniques enabled to overcome the restrictions of laser welding related to gap limitations, distortion of coverings, and vent-hole production in the liquified materials, through utilization of electric arc.

The idea behind the utilization of multiple thermal resources is applied in the domain of Metal Additive Manufacturing to enhance the process stableness by granting secondary energy (Qian et al., 2006). For instance, the use of laser to utilize the plasma arc deposition to figure out the newly developed Hybrid Manufacturing technique called laser plasma deposition fabrication (Zhang et al., 2006).

Zhang et al. (2018) introduced a laser utilization system to build thin-walled components of aluminium and evidenced the idea behind the control of height and non-uniformity of width of wall. Other researchers concentrated on the evaluation and analysis of the execution of the latest hybrid Additive Manufacturing techniques related to the deposition scheme and microstructures (Li et al., 2020)

Wu used a laser technology variation to build an aluminium component, which consisted of exceptionally good mechanical characteristics and microstructures. When it was compared with the conventional manufacturing techniques, it was found with small porosity and cracks (Wu et al., 2020).

The availability of a Hybrid Manufacturing method that combines turning or milling specifications with manufacturing parameters of metal additive to make ready-to-use metal components using a single clamp technique in the shortest delivery time (Lorenz et al., 2015; Merklein et al., 2016). When the manufacture of metal additives is hybridized with material removal in the final processing step, it is efficient while lowering production costs and material waste (Seow et al., 2019).

The overhanging notches are machined complex structures, and deep sections might be performed with conjugation of deposition of materials while routes of manufacturing (Luo & Frank, 2010). The latest research in metal-based hybrid additive manufacturing removing will only be initiated while expanding and merging in 2005 by customization of material removal techniques or by the Metal Additive Manufacturing (Kerschbaumer & Ernst, 2004).

On the identical line of improvement in hybrid multi-functioning systems, Valant and Kovacevic invented a robot system consisting of six axes to build metal components with laser-biased deposition and plasma-based operations (Kovacevic & Valant, 2006). Other researchers introduced a unique system consisting of milling and PA additive manufacturing techniques to fabricate an aero-engine impeller with double helix structure fabricated by nickel superior alloy (Xinhong et al., 2010).

Mazak invented the integral of L-DED with almost five-axis ability of machining or processing with integrex of 401 AM.

The initial Hybrid Manufacturing system depends upon the PBF technology called Lumex (Manogharan et al., 2015).

The system bonds the deposition of materials through LPBF by high-speed milling or it gains attraction with perfection towards outer contours, corrosive characters, and roughness of surfaces in moulds and dies (Ahn, 2011).

The struggle to enhance the flexibility of the deposition of metals, three-dimensional hybrid company offers the chance to segregate the LDED and GMA-DED into computer numeric control machining.

Another feature in metals hybrid Additive Manufacturing with regard to the forming procedures is the fabrication of pre-produced structures through Additive Manufacturing process to confirm the distortion-free stream and filling of dies with metals losing while the processes like forming, single level and small batch (Silva et al., 2017).

Metal forming equipment are produced through Additive Manufacturing to combine the several processes like:

- Combination with bulk forming.
- Combination with the sheet formation techniques.
- Combination with the joining by forming technique (Juncker et al., 2015).
- The initial utilization of rolling surface was with the sequential deposition of layers with WAAM techniques (Colegrove et al., 2013). When the rollers are placed opposite to the deposition torch, a different plan is developed for the utilization of pressure onto the layer of deposited material of WAAM techniques.
- It is demonstrated that the positive affection of rolling surfaces on the material characteristics is based on the deposited materials, because of the alteration of columnar structures. It has broaden the usage of metallic components in the Hybrid Additive Manufacturing with some portion by suggesting the integration of LDED with various portion techniques (Bamberg, 2012).
- In the advanced world, metal Hybrid Additive Manufacturing has broadened the application in various fields such as automotive, aerospace, and military industries, because it increases the characteristics of build components (Sealy, 2018).
- The extension of fatigue life is a result of the impact of compressive loads on delaying the initiation of fatigue cracks (Uzan et al., 2018).
- The combination of Metal Additive Manufacturing and hot rolling was introduced as a thermal and mechanical curing technique for big components of titanium alloy (Sokolov et al., 2020). The integration of Metal Additive Manufacturing with the hot forging by the applications of a custom WAAM torch was introduced by Duarte (Duarte et al., 2020). This acts like an optional solution in order to decrease the porosity, purification, and enhancement of micro-structures for the mechanical characteristics of the material deposited.
- Silva introduced the pioneering use of metal hybrid additive manufacturing in conjunction with extensive forming techniques (Silva et al., 2017).
- The investigation of the productivity of aluminium alloy that was deposited by WAAM was 'AA5083'. Silva performed various experimental and analytical analysis techniques on pre-prepared aluminium alloy with the deposition of cold-headed process.
- The determination of the high temperature distortion attitude of titanium alloy deposited through the LPBF in order to analyze the affection of strain rates on the development of microhardness, porosity, and structures.
- Papke Analyzed the productivity of stainless-steel deposition by LPBF with the medium of compressive testing (Papke et al., 2018).
- Bambach carried out the identical task on titanium alloy to indicate the practicability of the hybrid additive manufacturing paths with difficulty in processing materials (Bambach et al., 2020).

- Meiners repeated the feasibility of the addition of functional components by additive manufacturing on a partially finished component fabricated by the forging, regarding the depositing of material through LDED and WAAM in a T segment (Meiners et al., 2020). Michl applied the WAAM seating in a robotic structure to create annular preformance of deposition in mild steel with enhanced mass dispersion for the ring rolling (Michl et al., 2020).
- Two researchers presented the hybrid additive manufacturing paths in which the bending of sheet and the 3D specifications were joined by depositing with WAAM techniques.
- Analysis of the feasibility of putting on the characteristics on Ti6Al4V sheet just beginning and ending of LPBF process (Butzhammer et al., 2017).
- In the studies the impact of the stress state that it was produced as a result of bending operation on the strength connection between the features of deposited sheets. Rosenthal et al. (2019, 2020) presented an examination on the bending of flat monolithic sheets of Hastelloy X produced by process of additive manufacturing.
- Ahuja et al. (2015) were among the initial scientists to integrate deep drawing in metal additive manufacturing and added cylindrical features by utilizing LPBF on top of previously drawn titanium sheets.
- Two distinct routes of Hybrid Additive Manufacturing (HAM) have been evaluated by Bambach et al. (2017) which were based on the blend of deep drawing and tailored laser cladding by L-DED. The objective behind this was of saving weight, performance improvement, and lowering the possibility of extreme thinning.
- Schulte et al. (2020) approved the manufacture of deep drawn parts with a variety of teeth geometries that were formed with local material deposition by LPBF from orbital formed tailored blanks. By varying the quantity and geometry of three-dimensional cylindrical pins, Hafenecker et al. (2020) revised the elements of additively manufacturing on sheet formability.
- Recently, Pragana et al. (2020) investigated the WAAM process implied by incremental forming on a specific point of AISI 316L stainless steel sheets.
- Silva et al. (2019) employed additive manufacturing in combination with forming for creation of 'mortise-and-tenon' joints between two corresponding metal-polymer and metal-metal sheets. The above stated Hybrid Additive Manufacturing process was applied by Baptista et al. (2020) for connecting the hollow sections of aluminium profiles to the aggregate sheets for demonstrating the feasibility of this process in comparison to overlapped sheets.

1.7   CONCLUSION

The overview and background of Additive Manufacturing along with the hybridization of Additive Manufacturing processes has been consolidated in the last decades. Hybrid Manufacturing has been defined as the widespread technique to utilize the materials in the form of pellets, powder, tubes, and sheets. The new approach results in the overall monitoring of process mechanics while depositing raw materials compared with the characteristics of other traditional techniques.

The significance analysis on the hybridization of AM with the forming processes, it enhances the mechanical characteristics of additive deposited materials. The main outcomes of these studies are as follows:

- To enhance the utilization and to withstand the restriction of Additive Manufacturing regarding dimensional precision, rough surfaces, and productivity.
- To emphasize the fostering and flexibleness of new utilization with traditional manufacturing techniques.

REFERENCES

Ahn, D. G. (2011). Applications of laser assisted metal rapid tooling process to manufacture of molding & forming tools—State of the art. *International Journal of Precision Engineering and Manufacturing*, *12*(5), 925–938.

Ahuja, B., Schaub, A., Karg, M., Schmidt, R., Merklein, M., & Schmidt, M. (2015, March). High power laser beam melting of Ti-6Al-4V on formed sheet metal to achieve hybrid structures. In *Laser 3D Manufacturing II* (Vol. 9353, pp. 118–127).

Appuhamillage, G. A., Chartrain, N., Meenakshisundaram, V., Feller, K. D., Williams, C. B., & Long, T. E. (2019). 110th anniversary: Vat photopolymerization-based additive manufacturing: Current trends and future directions in materials design. *Industrial & Engineering Chemistry Research*, *58*(33), 15109–15118.

Araghi, B. T., Manco, G. L., Bambach, M., & Hirt, G. (2009). Investigation into a new hybrid forming process: Incremental sheet forming combined with stretch forming. *CIRP Annals*, *58*(1), 225–228.

Bambach, M. D., Bambach, M., Sviridov, A., & Weiss, S. (2017). New process chains involving additive manufacturing and metal forming–a chance for saving energy?. *Procedia Engineering*, *207*, 1176–1181.

Bambach, M., Sizova, I., Sydow, B., Hemes, S., & Meiners, F. (2020). Hybrid manufacturing of components from Ti-6Al-4V by metal forming and wire-arc additive manufacturing. *Journal of Materials Processing Technology*, *282*, 116689.

Bamberg, J., Hess, T., Hessert, R., & Satzger, W. (2012). Verfahren zum herstellen, reparieren oder austauschen eines bauteils mit verfestigen mittels druckbeaufschlagung. Mtu Aero Engines Gmbh, Munich, Germany, Patent No. WO, 2012152259, A1.

Baptista, R. J. S., Pragana, J. P. M., Bragança, I. M. F., Silva, C. M. A., Alves, L. M., & Martins, P. A. F. (2020). Joining aluminium profiles to composite sheets by additive manufacturing and forming. *Journal of Materials Processing Technology*, *279*, 116587.

Baufeld, B., Van der Biest, O., & Gault, R. (2010). Additive manufacturing of Ti–6Al–4V components by shaped metal deposition: Microstructure and mechanical properties. *Materials & Design*, *31*, 106–111. https://doi.org/10.1016/j.matdes.2009.11.032

Bhavar, V., Kattire, P., Patil, V., Khot, S., Gujar, K., Singh, R. (2014). A review on powder bed fusion technology of metal additive manufacturing. In *4th International Conference and Exhibition on Additive Manufacturing Technologies-AM-2014*, 1–2.

Butzhammer, L., Dubjella, P., Huber, F., Schaub, A., Aumüller, M., Baum, A., & Schmidt, M. (2017). Experimental investigation of a process chain combining sheet metal bending and laser beam melting of Ti-6Al-4V. In *Proceedings of World of Photonics Congress: Lasers in Manufacturing—LiM, Munich, Germany, June 26–29*.

Chu, W. S., Kim, C. S., Lee, H. T., Choi, J. O., Park, J. I., Song, J. H., & Ahn, S. H. (2014). Hybrid manufacturing in micro/nano scale: A review. *International Journal of Precision Engineering and Manufacturing-Green Technology*, *1*, 75–92.

Colegrove, P. A., Coules, H. E., Fairman, J., Martina, F., Kashoob, T., Mamash, H., & Cozzolino, L. D. (2013). Microstructure and residual stress improvement in wire and arc additively manufactured parts through high-pressure rolling. *Journal of Materials Processing Technology*, *213*(10), 1782–1791.https://doi.org/10.1016/j.jmatprotec.2013 .04.012

Crump, S. S. (1991). Fast, precise, safe prototypes with FDM. *ASME, PED*, *50*, 53–60. http:// www.3dhybridsolutions.com/

DebRoy, T., Wei, H. L., Zuback, J. S., Mukherjee, T., Elmer, J. W., Milewski, J. O., … Zhang, W. (2018). Additive manufacturing of metallic components–process, structure and properties. *Progress in Materials Science*, *92*, 112–224.

Derazkola, H. A., Khodabakhshi, F., & Simchi, A. (2020). Evaluation of a polymer-steel laminated sheet composite structure produced by friction stir additive manufacturing (FSAM) technology. *Polymer Testing*, *90*, 106690.https://doi.org/10.1016/j.polymertesting.2020.106690

Dickens, P. M., Pridham, M. S., Cobb, R. C., Gibson, I., & Dixon, G. (1992). Rapid prototyping using 3-D welding. In *1992 International Solid Freeform Fabrication Symposium*. http://hdl.handle.net/2152/64409

Duarte, V. R., Rodrigues, T. A., Schell, N., Miranda, R. M., Oliveira, J. P., & Santos, T. G. (2020). Hot forging wire and arc additive manufacturing (HF-WAAM). *Additive Manufacturing*, *35*, 101193.https://doi.org/10.1016/j.addma.2020.101193

Fuchs, J., Schneider, C., & Enzinger, N. (2018). Wire-based additive manufacturing using an electron beam as heat source. *Welding in the World*, *62*(2), 267–275. https://doi.org/ 10.1007/s40194-017-0537-7

Gonzalez-Gutierrez, J., Cano, S., Schuschnigg, S., Kukla, C., Sapkota, J., & Holzer, C. (2018). Additive manufacturing of metallic and ceramic components by the material extrusion of highly-filled polymers: A review and future perspectives. *Materials*, *11*(5), 840. https://doi.org/10.3390/ma11050840

Hull, C. W. (1990). *U.S. Patent No. 4,929,402*. Washington, DC: U.S. Patent and Trademark Office.

Junker, D., Hentschel, O., Schmidt, M., & Merklein, M. (2015). Qualification of laser based additive production for manufacturing of forging tools. In *MATEC Web of Conferences* (Vol. 21, p. 08010). EDP Sciences. https://doi.org/10.1051/matecconf/20152108010

Kalman, H., Tripathi, N. M., Gabrieli, O. G., & Portnikov, D. (2017). Phase diagrams for pneumatic and hydraulic conveying. In *18th International Conferences on Transport and Sedimentation of Solid Particles, T and S 2017* (pp. 145–152). Wydawnictwo Uniwersytetu Przyrodniczego we Wrocławiu.

Kerschbaumer, M., & Ernst, G. (2004, October). Hybrid manufacturing process for rapid high performance tooling combining high speed milling and laser cladding. In *Pacific International Conference on Applications of Lasers and Optics*. AIP Publishing.

Kodama, H. (1981). Automatic method for fabricating a three-dimensional plastic model with photo-hardening polymer. *Review of Scientific Instruments*, *52*(11), 1770–1773. https:// doi.org/10.1063/1.1136492

Körner, C. (2016). Additive manufacturing of metallic components by selective electron beam melting—A review. *International Materials Reviews*, *61*(5), 361–377.

Kovacevic, R., & Valant, M. E. (2006). *U.S. Patent No. 7,020,539*. Washington, DC: U.S. Patent and Trademark Office.

Kozak, J., & Rajurkar, K. P. (2000, September). Hybrid machining process evaluation and development. In *Proceedings of 2nd International Conference on Machining and Measurements of Sculptured Surfaces, Keynote Paper, Krakow* (pp. 501–536).

Larson, R. (1998). *U.S. Patent No. 5,786,562*. Washington, DC: U.S. Patent and Trademark Office.

Lauwers, B., Klocke, F., Klink, A., Tekkaya, A. E., Neugebauer, R., & Mcintosh, D. (2014). Hybrid processes in manufacturing. *CIRP annals*, *63*(2), 561–583.

Li, R., Wang, G., Ding, Y., Tang, S., Chen, X., Dai, F., … Zhang, H. (2020). Optimization of the geometry for the end lateral extension path strategy to fabricate intersections using laser and cold metal transfer hybrid additive manufacturing. *Additive Manufacturing*, *36*, 101546.

Lorenz, K. A., Jones, J. B., Wimpenny, D. I., & Jackson, M. R. (2015). A review of hybrid manufacturing. *Solid Freeform Fabrication Conference Proceedings*, *53*, 96–108.

Lumay, G., Tripathi, N. M., & Francqui, F. (2019). How to gain a full understanding of powder flow properties, and the benefits of doing so. *ONdrugDelivery*, *2019*, 42–47.

Luo, X., & Frank, M. C. (2010). A layer thickness algorithm for additive/subtractive rapid pattern manufacturing. *Rapid Prototyping Journal*, *16*(2), 100–115.https://doi.org/10.1108/13552541011025825

M Tripathi, N., & S Mallick, S. (2017). Pneumatic conveying of Fly Ash: Bend Models investigation. *Advanced Materials Proceedings*, *2*(8), 526–531.

Manogharan, G., Wysk, R., Harrysson, O., & Aman, R. (2015). AIMS–a metal additive-hybrid manufacturing system: System architecture and attributes. *Procedia Manufacturing*, *1*, 273–286. https://doi.org/10.1016/j.promfg.2015.09.021

Martina, F., Mehnen, J., Williams, S. W., Colegrove, P., & Wang, F. (2012). Investigation of the benefits of plasma deposition for the additive layer manufacture of Ti–6Al–4V. *Journal of Materials Processing Technology*, *212*(6), 1377–1386. https://doi.org/10.1016/ j.jmatprotec.2012.02.002

Meiners, F., Ihne, J., Jürgens, P., Hemes, S., Mathes, M., Sizova, I., & Weisheit, A. (2020). New hybrid manufacturing routes combining forging and additive manufacturing to efficiently produce high performance components from Ti-6Al-4V. *Procedia Manufacturing*, *47*, 261–267. https://doi.org/10.1016/j.promfg.2020.04.215

Merklein, M., Junker, D., Schaub, A., & Neubauer, F. (2016). Hybrid additive manufacturing technologies–an analysis regarding potentials and applications. *Physics Procedia*, *83*, 549–559. https://doi.org/10.1016/j.phpro.2016.08.057

Michl, D., Sydow, B., & Bambach, M. (2020). Ring rolling of pre-forms made by wire-arc additive manufacturing. *Procedia Manufacturing*, *47*, 342–348. https://doi.org/10.1016/j.promfg.2020.04.275

Murr, L. E., Martinez, E., Amato, K. N., Gaytan, S. M., Hernandez, J., Ramirez, D. A., … Wicker, R. B. (2012). Fabrication of metal and alloy components by additive manufacturing: Examples of 3D materials science. *Journal of Materials Research and technology*, *1*(1), 42–54.https://doi.org/10.1016/S2238

Nau, B., Roderburg, A., & Klocke, F. (2011). Ramp-up of hybrid manufacturing technologies. *CIRP Journal of Manufacturing Science and Technology*, *4*(3), 313–316. https:// doi.org/10.1016/j.cirpj.2011.04.003

Neveu, A., Lumay, G., Pillitteri, S., Monsuur, F., Pauly, T., Ribeyre, Q., Francqui, F., Vandewalle, N., & Tripathi, N. M. (2020a). Physical characterization of blends containing mesoporous particles with a focus on electrostatic properties. In *2020 AIChE Spring Meeting and 16th Global Congress on Process Safety*. AIChE.

Neveu, A., Tripathi, N. M., Rigo, O., Francqui, F., & Lumay, G. (2020b). Experimental investigation of spreadability of metal powders in recoating process. In *Proceedings - Euro PM2020 Congress and Exhibition (Proceedings - Euro PM2020 Congress and Exhibition)*. European Powder Metallurgy Association (EPMA).

Nielsen, C. V., & Martins, P. A. (2021). *Metal forming: Formability, Simulation, and Tool Design*. Academic Press.

Ning, Y., Wong, Y. S., & Fuh, J. Y. H. (2005). Effect and control of hatch length on material properties in the direct metal laser sintering process. *Proceedings of the Institution of Mechanical Engineers, Part B: Journal of Engineering Manufacture, 219*(1), 15–25. https://doi.org/10.1243/095440505X7957

Palanivel, S., & Mishra, R. S. (2017). Building without melting: A short review of friction-based additive manufacturing techniques. *International Journal of Additive and Subtractive Materials Manufacturing, 1*(1), 82–103. https://doi.org/10.1504/

Papke, T., Dubjella, P., Butzhammer, L., Huber, F., Petrunenko, O., Klose, D., & Merklein, M. (2018). Influence of a bending operation on the bonding strength for hybrid parts made of Ti-6Al-4V. *Procedia CIRP, 74*, 290–294.

Pragana, J. P., Pombinha, P., Duarte, V. R., Rodrigues, T. A., Oliveira, J. P., Bragança, I. M., Miranda, R., Coutinho, L., & Silva, C. M. (2020). Influence of processing parameters on the density of 316L stainless steel parts manufactured through laser powder bed fusion. *Proceedings of the Institution of Mechanical Engineers, Part B: Journal of Engineering Manufacture, 234*(9), 1246–1257. https://doi.org/10.1177/0954405420911768

Prinz, F. B., & Weiss, L. E. (1993). *U.S. Patent No. 5,207,371*. Washington, DC: U.S. Patent and Trademark Office.

Rajurkar, K. P., Zhu, D., McGeough, J. A., Kozak, J., & De Silva, A. (1999). New developments in electro-chemical machining. *CIRP Annals, 48*(2), 567–579. https://doi.org/10.1016/S0007-8506(07)63235-1

Ratsimba, A., Zerrouki, A., Tessier-Doyen, N., Nait-Ali, B., André, D., Duport, P., Neveu, A., Tripathi, N., Francqui, F., & Delaizir, G. (2021). Densification behaviour and three-dimensional printing of Y2O3 ceramic powder by selective laser sintering. *Ceramics International, 47*(6), 7465–7474. https://doi.org/10.1016/j.ceramint.2020.11.087

Rosenthal, S., Hahn, M., & Tekkaya, A. E. (2019). Simulation approach for three-point plastic bending of additively manufactured Hastelloy X sheets. *Procedia Manufacturing, 34*, 475–481. https://doi.org/10.1016/j.promfg.2019.06.201

Rosenthal, S., Maaß, F., Kamaliev, M., Hahn, M., Gies, S., & Tekkaya, A. E. (2020). Lightweight in automotive components by forming technology. *Automotive Innovation, 3*(3), 195–209. https://doi.org/10.1007/s42154-020-00103-3

Saboori, A., Gallo, D., Biamino, S., Fino, P., & Lombardi, M. (2017). An overview of additive manufacturing of titanium components by directed energy deposition: Microstructure and mechanical properties. *Applied Sciences, 7*(9), 883. https://doi.org/10.3390/app7090883

Sachs, E. M., Haggerty, J. S., Cima, M. J., Williams, P. A. (1993). *U.S. Patent No. 5,204. U.S. Patent and Trademark Office.* Washington, DC, p. 55.

Schulte, R., Papke, T., Lechner, M., & Merklein, M. (2020, November). Additive manufacturing of tailored blank for sheet-bulk metal forming processes. In *IOP Conference Series: Materials Science and Engineering* (Vol. 967, No. 1, p. 012034). IOP Publishing.

Sealy, M. P., Madireddy, G., Williams, R. E., Rao, P., & Toursangsaraki, M. (2018). Hybrid processes in additive manufacturing. *Journal of Manufacturing Science and Engineering, 140*(6), 060801. https://doi.org/10.1115/1.4038644.

Seow, C. E., Coules, H. E., Wu, G., Khan, R. H., Xu, X., & Williams, S. (2019). Wire+ Arc additively manufactured inconel 718: effect of post-deposition heat treatments on microstructure and tensile properties. *Materials & Design, 183*, 108157. https://doi.org/10.1016/j.matdes.2019.108157

Sharma, A., Babbar, A., Tian, Y., Pathri, B. P., Gupta, M., & Singh, R. (2022). Machining of ceramic materials: A state of the art review. *International Journal on Interactive Design and Manufacturing (IJIDeM)*, 1–21. https://doi.org/10.1007/s12008-022-01016-7

Sharma, A., & Jain, V. (2020). Experimental investigation of cutting temperature during drilling of float glass specimen. In *IOP Conference Series: Materials Science and Engineering* (Vol. 715, No. 1, p. 012050). IOP Publishing.

Sharma, A., Jain, V., & Gupta, D. (2018). Characterization of chipping and tool wear during drilling of float glass using rotary ultrasonic machining. *Measurement*, *128*. 254–263.

Sharma, A., Jain, V., & Gupta, D. (2019a). Tool wear analysis while creating blind holes on float glass using conventional drilling: A multi-shaped tools study. In *Advances in Manufacturing Processes: Select Proceedings of ICEMMM 2018* (pp. 175–183). Springer Singapore.

Sharma, A., Jain, V., & Gupta, D. (2019b). Comparative analysis of chipping mechanics of float glass during rotary ultrasonic drilling and conventional drilling: For multi-shaped tools. *Machining Science and Technology*, *23*(4), 547–568.

Sharma, A., Jain, V., & Gupta, D. (2019c). Multi-shaped tool wear study during rotary ultrasonic drilling and conventional drilling for amorphous solid. *Proceedings of the Institution of Mechanical Engineers, Part E: Journal of Process Mechanical Engineering*, *233*(3), 551–560.

Sharma, A., Jain, V., & Gupta, D. (2021a). Effect of pre and post tempering on hole quality of float glass specimen: For rotary ultrasonic and conventional drilling. *Silicon*, *13*, 2029–2039.

Sharma, A., Jain, V., & Gupta, D. (2021b). Mathematical approach on chipping volume estimation generated during rotary ultrasonic drilling for float glass. *Proceedings of the National Academy of Sciences, India Section A: Physical Sciences*, *92*, 285–291.

Sharma, A., Jain, V., Gupta, D., & Babbar, A. (2020). A Review Study on Miniaturization, *Advanced Manufacturing and Processing Technology*, First edition. Boca Raton, FL: CRC Press, 111–131.

Shellabear, M., & Nyrhilä, O. (2004). DMLS-Development history and state of the art. *Laser Assisted Netshape Engineering 4*, *Proceedings of the 4th LANE*, 21–24.

Silva, C. M. A., Bragança, I. M. F., Cabrita, A., Quintino, L., & Martins, P. A. F. (2017). Formability of a wire arc deposited aluminium alloy. *Journal of the Brazilian Society of Mechanical Sciences and Engineering*, *39*, 4059–4068. https://doi.org/10.1007/s40430-017-0864-z

Silva, D. F., Braganca, I. M., Silva, C. M., Alves, L. M., & Martins, P. A. (2019). Joining by forming of additive manufactured 'mortise-and-tenon'joints. *Proceedings of the Institution of Mechanical Engineers, Part B: Journal of Engineering Manufacture*, *233*(1), 166–173. https://doi.org/10.1177/0954405417720954

Sokolov, P., Aleshchenko, A., Koshmin, A., Cheverikin, V., Petrovskiy, P., Travyanov, A., & Sova, A. (2020). Effect of hot rolling on structure and mechanical properties of Ti-6Al-4V alloy parts produced by direct laser deposition. *The International Journal of Advanced Manufacturing Technology*, *107*, 1595–1603. https://doi.org/10.1007/s00170-020-05132-0

Tripathi, N. M., Francqui, F., & Lumay, G. (2020). Influence of relative air humidity on the flow property of fine powders. In *Third International Conference on Powder, Granule and Bulk Solids: Innovations and Applications PGBSIA 2020 February 26–28, 2020* (p. 63).

Tripathi, N. M., Francqui, F., Pirenne, T., & Lumay, G. (2019). Measuring food powders electrical properties as a result of anti-static content. In *2019 AIChE Annual Meeting*. American Institute of Chemical Engineers.

Tripathi, N. M., Levy, A., & Kalman, H. (2016). Initial acceleration pressure drop in dilute phase pneumatic conveying system. In *Powder, Granule and Bulk Solids: Innovations and Applications Conference*.

Tripathi, N. M., & Mallick, S. S. (2014). *An Investigation into Pressure Drop Across Bends for Fluidised Densephase Pneumatic Conveying Systems* (Doctoral dissertation).

Uzan, N. E., Ramati, S., Shneck, R., Frage, N., & Yeheskel, O. (2018). On the effect of shot-peening on fatigue resistance of AlSi10Mg specimens fabricated by additive manufacturing using selective laser melting (AM-SLM). *Additive Manufacturing*, *21*, 458–464. https://doi.org/10.1016/j.addma.2018.03.030

Williams, S. W., Martina, F., Addison, A. C., Ding, J., Pardal, G., & Colegrove, P. (2016). Wire þ arc additive manufacturing. *Materials Science and Technology*, *32* (7), 641–647. https://doi.org/ 10.1179/1743284715Y.0000000073

Wohlers, T. (2017). Desktop metal: A rising star of metal AM targets speed, cost and high-volume production. *Metal AM*, *3*(2), 89–94. http://www.metal-am.com /wp-content/upl oads/sites/4/2017/06/MAGAZINE-Metal-AM-Summer-2017-PDF-sp.pdf.

Wu, B., Pan, Z., Ding, D., Cuiuri, D., Li, H., Xu, J., & Norrish, J. (2018). A review of the wire arc additive manufacturing of metals: Properties, defects and quality improvement. *Journal of Manufacturing Processes*, *35*, 127–139. https://doi.org/10.1016/j.jmapro .2018.08.001

Wu, D., Liu, D., Niu, F., Miao, Q., Zhao, K., Tang, B., … Ma, G. (2020). Al–Cu alloy fabricated by novel laser-tungsten inert gas hybrid additive manufacturing. *Additive Manufacturing*, *32*, 100954. https://doi.org/10.1016/j.addma.2019.100954

Xinhong, X., Haiou, Z., Guilan, W., & Guoxian, W. (2010). Hybrid plasma deposition and milling for an aeroengine double helix integral impeller made of superalloy. *Robotics and Computer-Integrated Manufacturing*, *26*(4), 291–295. https://doi.org/10.1016/

Yap, Y. L., Wang, C., Sing, S. L., Dikshit, V., Yeong, W. Y., & Wei, J. (2017). Material jetting additive manufacturing: An experimental study using designed metrological benchmarks. *Precision Engineering*, *50*, 275–285. https://doi.org/10.1016/ j.precisioneng.2017.05.015

Zhang, H. O., Qian, Y., Wang, G., & Zheng, Q. (2006). The characteristics of arc beam shaping in hybrid plasma and laser deposition manufacturing. *Science in China Series E*, *49*, 238–247. https://doi.org/10.1007/s11431-006-0238-8

Zhang, Z., Sun, C., Xu, X., & Liu, L. (2018). Surface quality and forming characteristics of thin-wall aluminium alloy parts manufactured by laser assisted MIG arc additive manufacturing. *International Journal of Lightweight Materials and Manufacture*, *1*(2), 89–95. https://doi.org/10.1016/j.ijlmm.2018.03.005

Zhu, Z., Dhokia, V. G., Nassehi, A., & Newman, S. T. (2013). A review of hybrid manufacturing processes–state of the art and future perspectives. *International Journal of Computer Integrated Manufacturing*, *26*(7), 596–615. https://doi.org/10.1080 /0951192X.2012.749530

Ziaee, M., & Crane, N. B. (2019). Binder jetting: A review of process, materials, and methods. *Additive Manufacturing*, *28*, 781–801. https://doi.org/10.1016/j.addma.2019.05.031

2 Integrating Titanium Coating with 3D-Printed GFRP Panels

An Innovative Approach to Harnessing Composite Strengths in Engineering Applications

Satishkumar P., Ravi Kumar, Naveen Mani Tripathi, Ankit Sharma, Dhaval Jaydevkumar Desai, Ashok Verma Natarajan N., and Atul Babbar

2.1 INTRODUCTION

Composite materials, like Polymer Matrix Composites (PMCs), are gaining popularity due to their anisotropic structure. Their properties can be completely tailored to suit the requirements. There is a growing need for engineered materials that can meet the wide range of requirements that arise based on their uses. In the recent years, numerous engineering disciplines have shown keen interest in PMCs due to the material's potential that allows it to be used in high-performance, lightweight structures. According to studies (Manikandan et al. 2023; Muruganandhan et al. 2023), enhancing Polymer Matrix Composites (PMCs) through metallization is a leading strategy to overcome their primary limitations, such as their inadequate surface properties and high thermal vulnerability. Surface metallization offers PMCs advantages like improved thermal and electrical conductivity, enhanced erosion resistance, and reliable lightning protection for aviation applications. While several methods exist for PMC surface metallization, Cold Spray (CS) technology is particularly notable. It offers the advantage of applying thick coatings without undermining the substrate's integrity (Ramesh et al. 2022). In CS technology, micron-sized metallic powders (ranging from 10 m to 100 m in diameter) are deposited using kinetic energy without being melted, thereby maintaining their inherent qualities. These powders are

DOI: 10.1201/9781003428862-2

accelerated to supersonic speed and passed through de Laval nozzle, using gases like air, nitrogen, or helium. Upon hitting the surface at a speed ranging from 200 m/s to 1200 m/s, the powders experience intense plastic deformation, bonding cohesively to form a uniform coating (Kang et al. 2022). The key factors influencing the Cold Spray process include the Standoff Distance (SoD), temperature, pressure, and choice of gas. It's widely acknowledged that by diligently controlling these parameters, a successful deposition can be achieved (Radojičić et al. 2023).

Amid the growing excitement around using the Cold Spray technique for the application of metallic powders on PMCs, significant gaps in knowledge exist. Specifically, the exact bonding process between the powder and the underlying material remains elusive (Horňáček et al. 2022). Moreover, definitive guidelines on how to secure optimal coating quality are yet to be defined. In order to find the optimal CS condition, researchers often experiment with various combinations of powder and substrate and experiment on different values for the parameters of the deposition process (Joch et al. 2023; Sitotaw et al. 2022; Miri et al. 2022). To metallize PEEK polymer substrates, for instance, the authors sprayed titanium (Ti) particles under 41 bar of gas pressure from a 75-mm standoff. As indicated by researchers (Borg, Kiss, & Rochman 2021), there's a variation in the preferred parameters for coating similar substrates with Al-based particles, applying gas pressures between 3 bar and5 bar and SoDs ranging from 25 mm to 125 mm. This disparity underscores the diverse perspectives within academic circles on the optimal process parameters and tools for achieving quality deposition. As PMC metallization via CS is just in the initial stages, there are still many unanswered questions that need to be investigated (Grujovic et al. 2017; Araya-Calvo et al. 2018; Jia, Wang, & Liu 2023).

Recent research (Imaeda et al. 2021) has demonstrated the importance of paying close attention to the composite lay-up sequence and stratification in order to achieve an efficient deposition. Substrate stratification, like Standoff Distance and inlet gas pressure, is an important process parameter that must be taken into account. The depth of the matrix layer applied to the reinforced polymer's surface plays a role in particle deformation and subsequently influences coating uniformity. Research indicates that the Cold Spray technique can successfully deposit steel 316L powder onto composite substrates (Pernica et al. 2023; Ichihara & Ueda 2022). Such discoveries point to the potential for depositing harder materials, provided the correct Cold Spray parameters matrix layer thickness and lay-up sequence are employed. This insight underscores the idea that to achieve premium metallic coatings on PMCs, customizing the composite substrates to align with specific CS guidelines may be necessary, highlighting the potential for tailored production. Certain other research studies also pointed at an innovative approach to advancement in manufacturing processes (Sharma et al., 2022; Sharma & Jain, 2020; Sharma et al., 2018; Sharma et al., 2019a; Sharma et al., 2019b; Sharma et al., 2019c; Sharma et al., 2021a; Sharma et al., 2021b; Sharma et al., 2020; Sharma et al., 2022; Kalman et al., 2017; Lumay et al., 2019; Tripathi & Mallick, 2017; Neveu et al., 2020a; Neveu et al., 2020b; Ratsimba et al., 2021; Tripathi et al., 2020; Tripathi et al., 2019; Tripathi et al., 2016; Tripathi & Mallick, 2014).

There is a wide variety of methods available for making specialized substrates. With their low production costs and high level of automation, AM techniques stand out as the best option for obtaining custom composites by altering the stratification and sequence. Matrix and reinforcements can be deposited in a variety of ways, allowing for a wide range of surface morphologies to be created with AM (Kohn et al. 2021).

Fused Filament Fabrication (FFF) is a mature technology that can manufacture thermoplastic-based composites. Matrix and long fibres can be printed using this technology with a variety of deposition strategies and printing parameters due to the use of a double mobile nozzle (Figure 2.1) (Al Rashid & Koç 2021). In this technique, 3D objects are built layer by layer from polymer filaments by feeding them into a heated nozzle, which then extrudes the filaments in accordance with 3D CAD models. The bond between extruded polymers was vital in producing flawless objects. Their mechanical properties should be on par with those achieved through traditional extrusion and injection moulding techniques (Lee et al. 2023; Papa et al. 2022). A unique nozzle lays down the composite fibre in a continuous strand for every potential layer, with the flexibility to adjust the length of the reinforcing fibres, including glass, fibre glass, or Kevlar.

The versatility of Fused Filament Fabrication (FFF) technology allows for its integration with Cold Spray technology, offering tailored PMC substrates fit for specific CS applications. Recent research (Prajapati, Dave, and Raval 2020) indicates that the surface texture of 3D-printed panels, which can be manipulated during manufacturing, plays a pivotal role in determining the characteristics of the CS coatings, validating the combined FFF–CS approach. Specifically, they demonstrated the importance of powder particles being nestled between the paths created by consecutive polymer tracks and highlighted the proportionality between the distances in the texturing valleys and the diameter of the powder particles.

Despite these initial findings, the topic requires further exploration. The adhesive qualities of CSed metal coatings on PMCs have only been minimally covered in the existing literature (Vaško et al. 2020; Papa et al. 2021). To the best of the authors' understanding, none has drawn connections between adhesion test results, CS printing strategy, and the process parameters. This study seeks to elucidate the features of hybrid 3D-printed-CSed structures, focusing on refining the Cold Spray deposition

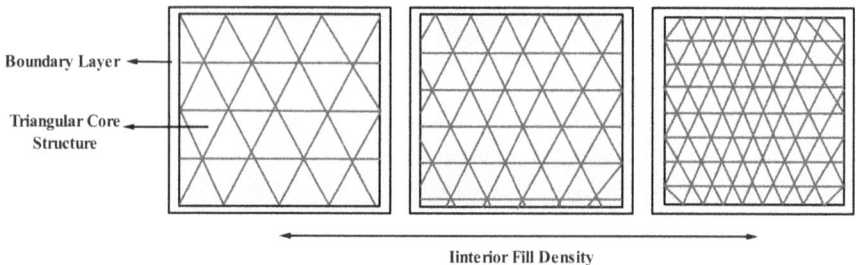

FIGURE 2.1 Schematic view of matrix printing of different interior fill density.

process and delving into how the manufacturing strategy of the panels influences the entire process (Rao et al. 2023).

To demonstrate the effect of the fibre on the deposition, 3D-printed unreinforced panels made of Onyx and long glass fibre-reinforced panels using a variety of stratification strategies were used. In 3D-printed laminates, Onyx, an innovative nylon-based polymer infused with short glass fibres, serves as the primary matrix material. Both the pure Onyx and reinforced panel types in the sample set were manufactured with a matrix that varied in infill density and layer thickness across the board. The filler percentages of 20%, 40%, and 60% were specifically considered in this exercise. Low-Pressure Cold Spray (LPCS) was used to metalize each laminate using the same CS process parameters established through prior studies. The metallization of PMCs made use of titanium powders on the micron scale. Final products were examined with optical, SEM, and confocal microscopes. After the metallization process, the coating height, particle splat size, distribution, and surface coverage of each specimen was measured. Each coating's adhesion strength was evaluated and compared to the findings from microscopic examinations.

In conclusion, our study examined the influence of infill density, surface layer thickness, and the inclusion of reinforcing fibres within the polymer on the CS deposition from 3D-printed substrates.

2.2   MATERIALS AND METHODS

The Markforged X7 3D printer was used to create the composite laminates using FFF technology. This printer has a dual nozzle that can extrude both matrix filaments and fibres. A built-in laser micrometre mapped the surface of the process plate and allowed for extremely precise calibration of the nozzle's height (1 m). Onyx was used for the matrix in these laminates, and long glass fibres were used for reinforcement. Thermoplastic on the base, Onyx is a nylon/glass fibre blend. Onyx has a temperature of 140°C at which it begins to reflect heat. Better performance is ensured by the presence of short fibres in comparison to unreinforced nylon. To be more specific, they guarantee improved dimensional stability by lowering thermal deformations by altering the nylon's behaviour upon cooling.

Long glass fibres that are printable were used, with a thin nylon film wrapped around each packet of fibres to ensure the best possible adhesion to the matrix. To enable the polymer's melting during printing, the extrusion nozzle's temperature was maintained at 260°C. The average extrusion speeds were 2.39 cm^3/hr for fibre and 6.90 cm^3/hr for Onyx layers.

Three different filling percentages 20%, 40%, and 60% were considered for all the samples, meanwhile, 1, 2, or 3 layers of only Onyx were printed on the surfaces of the panel to reap the benefits of the resultant surface polishing. Reinforced samples with a 0° and 90° symmetrical layout were fabricated by alternating matrix and fibre layers in accordance with the lay-up sequences (Figure 2.2).

Each matrix layer's boundaries must be established by the nozzle before a triangle pattern can be established for the matrix core structure. In total, 18 unique samples were 3D printed, as shown in Table 2.1. The table also includes a summary

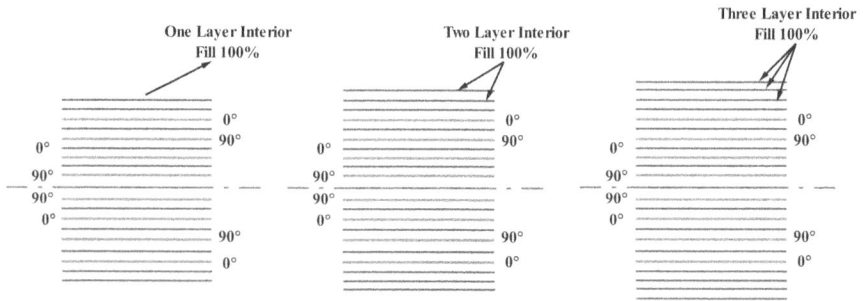

FIGURE 2.2 A straightforward plan for a 0° and 90° symmetric arrangement.

TABLE 2.1
Manufacturing 3D-Printed Samples by Fused Filament Fabrication Technique

No. of Specimens	Specimen Code for Panels	Number of Superficial Layers	Density of the Interior Fill [%]	Layer Thickness [mm]	Panel Thickness [mm]
1	S1	1	20	0.125	2.0
2	S2	1	40	0.125	2.0
3	S3	1	60	0.125	2.0
4	S4	2	20	0.125	2.25
5	S5	2	40	0.125	2.25
6	S6	2	60	0.125	2.25
7	S7	3	20	0.125	2.5
8	S8	3	40	0.125	2.5
9	S9	3	60	0.125	2.5
10	S10	1	20	0.125	2.0
11	S11	1	40	0.125	2.0
12	S12	1	60	0.125	2.0
13	S13	2	20	0.125	2.25
14	S14	2	40	0.125	2.25
15	S15	2	60	0.125	2.25
16	S16	3	20	0.125	2.5
17	S17	3	40	0.125	2.5
18	S18	3	60	0.125	2.5

of the most salient geometric features of each specimen. The outermost layers of the 80-mm square panels can be 2.0 mm or 2.5 mm thick. Each layer of the stratified panel has a thickness of around 0.130 mm, based on the profile of the heating.

2.2.1 USE OF COLD SPRAY FOR METALLIZATION

For CSD, titanium powders (micron-sized) were utilized in conjunction with air as the carrier gas in a Dycomet 423 low-pressure apparatus. For optimal deposition,

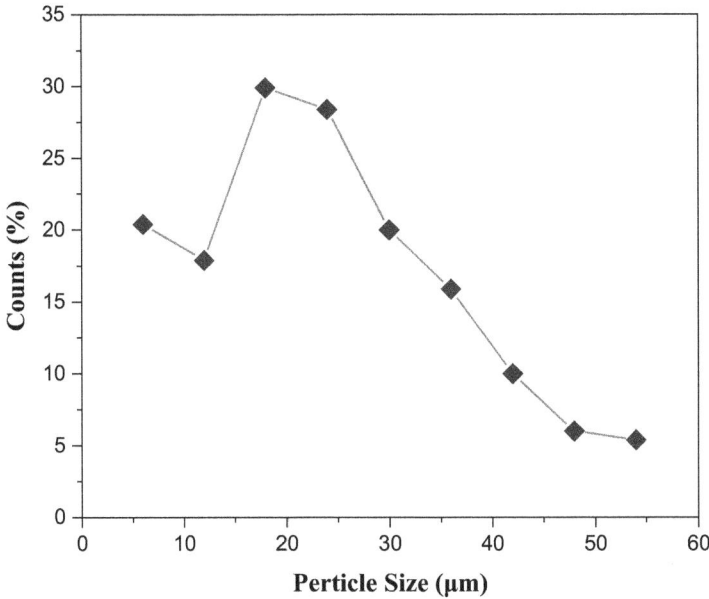

FIGURE 2.3   A powder size distribution.

research suggests using spherical particles with a mean diameter of 20 m. The selected powders' macroscopic appearance and size distribution are shown in Figure 2.3.

The spray cannon was held steady by the HIGH-Z S-400/T CNC-Technik robot, resulting in an even coating. For consistency, each substrate received three coated tracks from a single gun pass under consistent Cold Spray process conditions. The ideal Cold Spray process parameters were determined through a mix of initial tests and findings from related studies. Specifically, the inlet gas had a pressure of 6 bar and a temperature of 200°C. The Standoff Distance (SoD) was consistently set at 45 mm during the deposition.

It is important to note the role of carrier gas pressure in these settings: a greater pressure propels the particles more deeply into the polymer. However, excessive gas pressure might lead to erosion. The gas temperature, which was set in relation to the transition temperature of nylon, ensures effective deposition. Researchers (Dharmalingam et al. 2023) suggest that optimal deposition occurs when the incoming gas temperature exceeds the substrate's glass transition temperature, yet stays below its melting point to avert degradation.

In Figure 2.2, the grey stripes represent the Cold Spray method's single-track coatings. Both the pure Onyx and the fibre-reinforced samples have a distinguishing thin outer matrix layer.

2.2.2   PANELS CHARACTERIZATION

Coating height and surface morphology were assessed using confocal microscopy techniques, allowing us to make connections between panel manufacturing processes and CS deposition mechanisms. LeicaMap software was used to perform topographic analysis in accordance with the ISO 25178-2:2012 standard after a three-dimensional surface was generated from a randomly chosen specimen area. Peak height (Sp), valley depth (Sv), and surface height (Hs) were the primary metrics considered (Sz). Specifically, as indicated in the literature (Li et al. 2022), Sz was used as a metric to gauge the coating's height. This consideration is essential when interpreting and discussing the findings.

The Cold Spray process clearly deformed the substrate in the unreinforced samples, while the reinforced samples exhibited no such deformation due to the fibres' obvious stiffening effect. The distortion was quantified using the following formula. This was done to analyze the influence of Onyx interior fill approach and thickness of the matrix (Eq. 2.1) (Pace et al. 2023).

$$\delta = \frac{d - t - d_0}{d_0}\%$$
(2.1)

A Hitachi TM 2000 Scanning Electronic Microscope was used to take a closer look at the coatings on the laminates. The pixel threshold of the corresponding SEM images was analyzed, and a binary mask of the acquired surface was created to evaluate the latter. A size filter was then used as a digital sieve to separate the blobs. The six various diameter ranges were analyzed to calculate the percentage of surface area occupied by particles up to 60 μm in diameter. The adhesiveness was measured with an ASTM D4541 standard and a PosiTest ATM. Cyanoacrylate glue was used to attach 10-mm dia titanium dollies to the cold-sprayed deposits' exposed surfaces. A drill bit was used to chip away at the coating and the adhesive that had accumulated around the dolly. The amount of coating removed from the dolly surfaces was measured using an optical microscope to determine the test's success.

2.3   RESULTS

To explore the impact of surface layer thickness on CS deposition, panels were produced with one, two, or three matrix layers on their outer surfaces, employing a 45° pattern.

A distinct surface morphology where the underlying triangular pattern becomes evident, especially when a single matrix layer with a 45° pattern is applied to the surface. As can be seen in Figure 2.4, the triangle pattern tends to vanish as more layers are printed on the surface. Reinforced panels, which are not shown here for brevity's sake, exhibit the same behaviour. As a result, the Cold Spray deposition would be affected in different ways depending on the surface morphology into which the particles were impinging. The thickness of the layer and surface pattern both play a role in determining where the particles are deposited.

Table 2.1 encompasses the results from confocal analyses on all coated samples. To be concise, the surface morphology of pure Onyx panels, where only the exterior surfaces have a matrix layer, more infill (here, from 20% to 60%) causes the

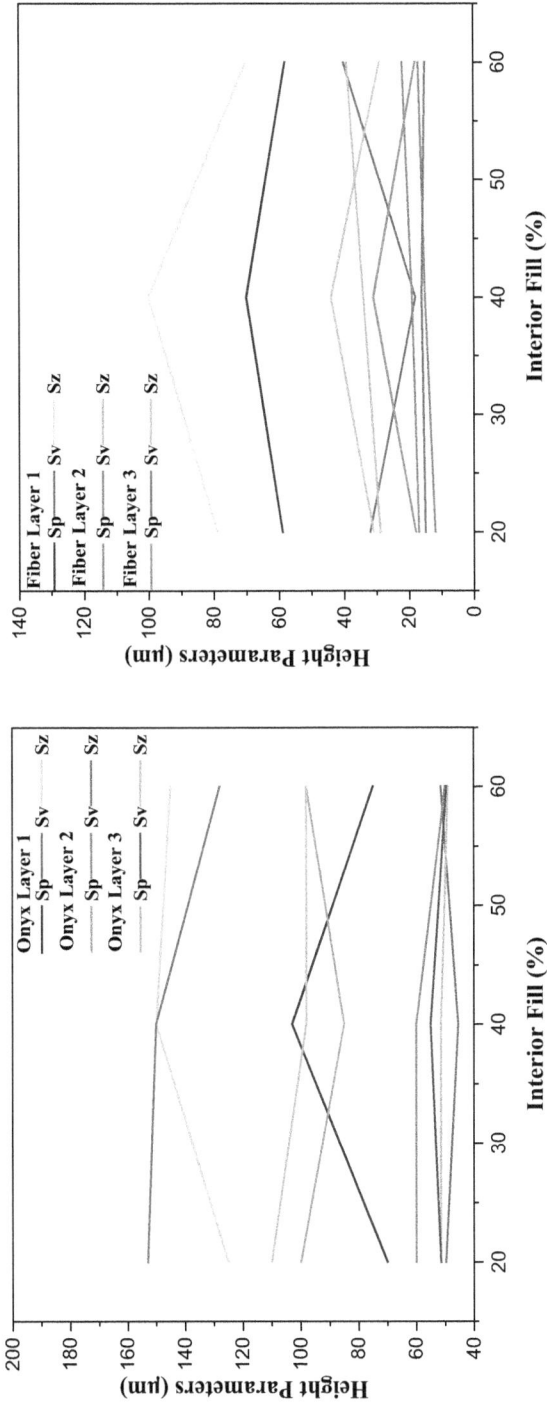

FIGURE 2.4 Height parameters from confocal analyses.

highlighted triangular structures to contract, leaving less space between neighbouring valleys.

The confocal analyses used to derive the height parameters are shown graphically in Figure 2.4. As depicted in the image, the parameters for the FRP are less for pure Onyx samples, irrespective of interior fill density. This indicates that the fibre-reinforced panels have a greater coating height (Sz) than the Onyx samples. According to these findings, particles deform more when striking reinforced or stiffer substrates.

2.3.1 SUBSTRATE DISTORTION

The Cold Sprayed deposits have distorted all the samples, making some of them concave and others convex. Glass-reinforced Onyx panels did not show signs of substrate distortion, and so they have not been included in this study. This discrepancy between the two substrate types is likely attributable to the Onyx panels' lower stiffening, which results in greater thermal deflection during the deposition.

Equation 2.1 is used to determine the substrate distortion, which is depicted graphically in Figure 2.5. Both matrix interior fill and outer substrate layers can influence the end product. As the count of surface layers rises from one to three, distortion diminishes, ranging from a peak of 225% to a trough of 65%. It is observed that escalating the matrix infill from 20% to 60% also lessens the substrate warping. With a 20% infill and just a single superficial matrix layer, substrate distortion can spike up to 225%.

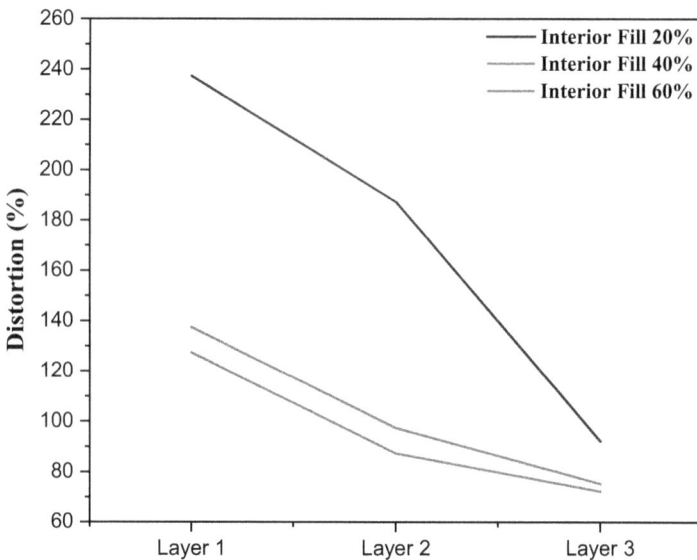

FIGURE 2.5 Outcomes of substrate distortion of pure Onyx panels.

Intriguingly, as the quantity of the outer layers amplifies, the variance due to the matrix infill percentage starts to level out. On the other hand, those with three exterior layers exhibit a much narrower distortion bracket (60–70%), showing little to no sway from the matrix infill. This pattern insinuates that the primary determinant of substrate distortion is the thickness of the superficial matrix. The diminishing impact of the matrix infill, as external layers multiply hinting at substrate reinforcement, further corroborates this observation.

The coatings derived from fibre-reinforced specimens manifest a higher degree of homogeneity. This uniformity can be attributed to the augmented deformation of the particles, a direct consequence of the fibres' stiffening effect. This phenomenon is vividly illustrated in Figure 2.6, which shows the variation in splat size for the particles. The enhanced rigidity intrinsic to these panels encourages greater particle deformation, leading to these observations. The splat size distribution analysis verifies this as well (Figure 2.7). Larger particles are present and cover the substrate surface through deformation, therefore the fibre-reinforced panels have a more uniform size distribution as compared to S4 panels.

2.3.2 BONDING STRENGTH

The outcomes of tests for bonding strength are depicted in Figure. 2.8. As can be seen from the data, the adhesion strength of Cold Sprayed coatings on pure Onyx samples is significantly higher than that of glass-reinforced Onyx samples. These numbers are comparable with those given for metallic coatings on polymeric substrates in the previous studies (Prajapati, Dave, & Raval 2023; 2021).

Coatings applied to pure Onyx panels have an adhesion strength between 3 MPa and 5.8 MPa, as shown in Figure 2.8. When holding the number of superficial layers constant, a rise in matrix infill leads to improved coating adhesion. Yet, akin to the pattern observed with substrate distortion, the variation in adhesion strength values among the panels (20%, 40%, and 60%) diminishes as the count of superficial matrix layers escalates. This suggests that while the matrix infill is influential, the number of superficial layers also plays a significant role in determining adhesion strength.

As can be seen in Figure 2.8, it is evident that carbon-reinforced Onyx substrates achieve an adhesion strength range between 2.6 MPa and 4.2 MPa. This suggests that the variation in adhesion strength is not as pronounced for these substrates. Furthermore, at a constant number of surface layers, the adhesion strength sees an uptick with an increment in the matrix infill. This underscores the nuanced relationship between material reinforcement, matrix infill, and surface layers in determining adhesion quality.

The effectiveness of the adhesion tests has been studied by optically analyzing coating particles that have remained adhered to the dolly surface. Three distinct fracture surface morphologies were identified: to begin, an AlSi10Mg coating, an Onyx base, and glue or other contaminants. To keep things simple, Figure 2.9 only shows the results achieved with 100% Onyx samples.

Figure 2.9 provides a clear indication that the Al coating has achieved complete coverage over the dollies. This thorough coverage suggests a consistent and effective application of the coating process. Furthermore, the observation that the bond

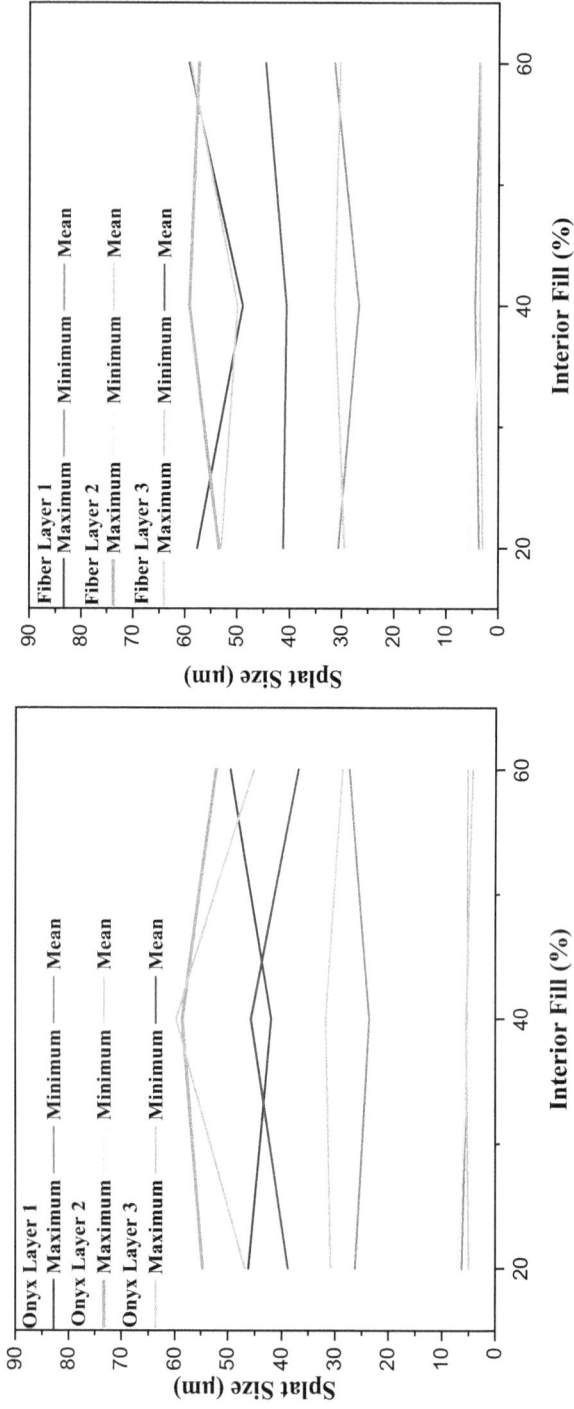

FIGURE 2.6   Particles splat size.

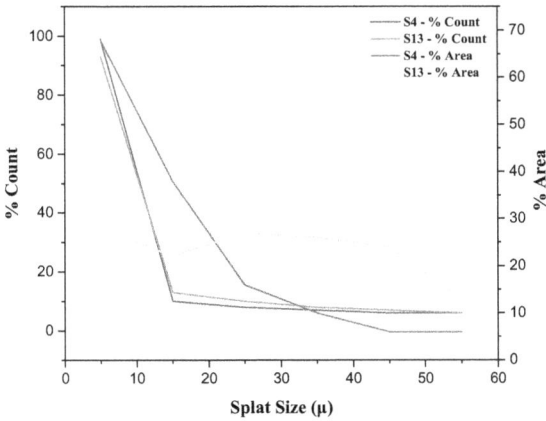

FIGURE 2.7 Distribution analysis of splat size.

strength between the Onyx substrate and the Ti coated is lower compared to the bond strength between coating and glue which reinforces the accuracy and effectiveness of the testing procedure. This implies that any failure during the test is more likely to occur between the Onyx and the Al coating, rather than between the glue and the coating.

2.4 DISCUSSION

Previous studies (Naik, Thakur, & Salunkhe 2023; Ahmadifar et al. 2022) have underscored the importance of the substrate's manufacturing method in influencing the outcomes of the Cold Spray deposition process. This study delves deeper into this aspect by considering several key factors. Firstly, the density of the infill fibres used to reinforce polymer panels is critically examined. Secondly, the thickness of the topmost layer and its effects on the Cold Spray deposition are studied. Both of these factors are crucial in understanding the nuances of the deposition process and how they might influence the final product's properties and performance.

2.4.1 EFFECT OF FIBRES

Figure 2.10 clearly indicates the influence of reinforcement on the average coating height. Reinforcement in the form of fibres added to the Onyx panels appears to drastically reduce the height of coating. This is attributed to two significant phenomena: distortion of sprayed particles and their subsequent penetration into the substrate. A lesser coating height on reinforced panels suggests that the sprayed particles either deform more or penetrate deeper into these panels as compared to the unreinforced ones.

The greater deformation of sprayed particles on reinforced panels is corroborated by Figure 2.10. This figure depicts that a significantly higher percentage of particles deform upon impacting reinforced panels. In other words, they expand beyond their

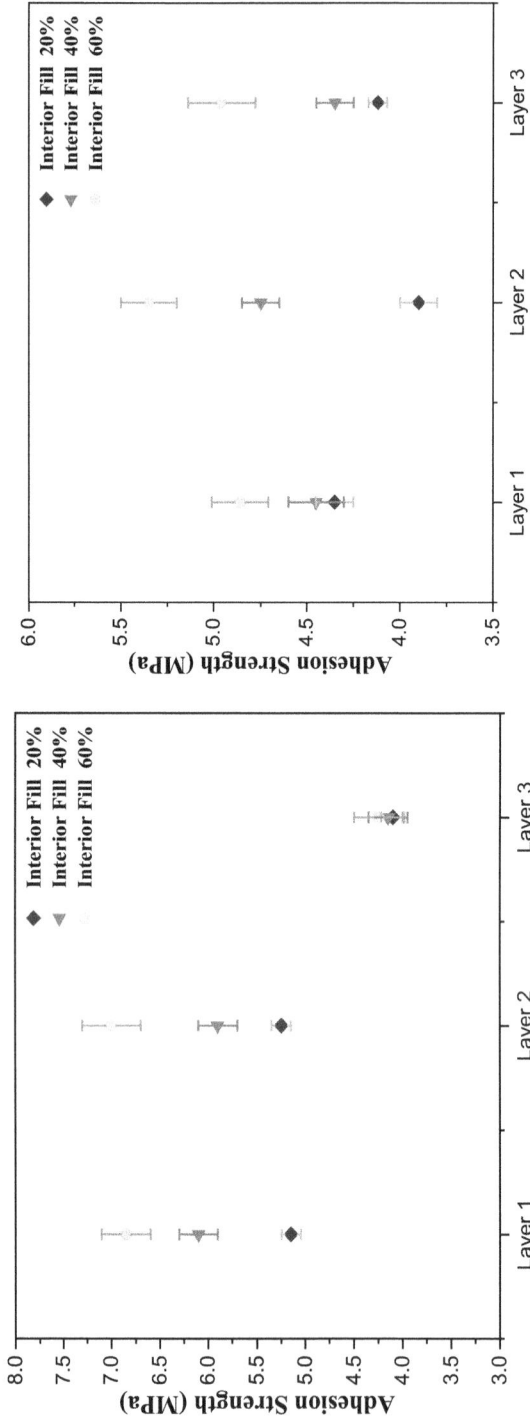

FIGURE 2.8 Outcomes of bonding strength tests for reinforced and pure Onyx specimens.

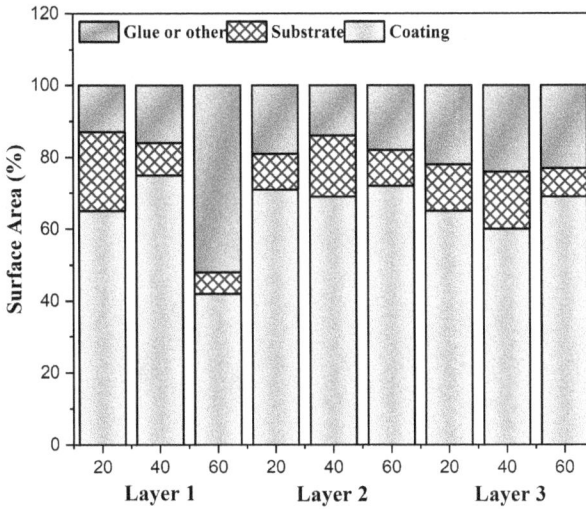

FIGURE 2.9 Area fractions for pure Onyx samples.

initial diameter upon impact. This is indicative of a greater level of deformation during the spraying process. Such behaviour is typically expected when particles impact a stiffer surface. Reinforcement, by its nature, increases the stiffness of materials. Moreover, the micrographs showing cross-sectional views of the specimens provide direct visual evidence of the penetration depth of the sprayed coatings.

Figure 2.10 shows that when coating is applied to reinforced panels, the resulting coating is less adhered to the substrate than when coating is applied to unreinforced panels. This finding contrasts with previous research which found that metallic coatings on metallic substrates exhibit greater adhesion due to their ability to withstand greater deformation. Therefore, increasing adhesion requires encouraging deeper particle penetration into substrates, while particle deformation appears to contribute less to adhesion. Overall, the presence of the fibres reduces the coating's adhesion to the substrate, but it does result in more uniform coatings.

2.4.2 IMPACT OF THE INTERIOR FILL DENSITY

The mechanisms in charge of metallic particle adhesion to 3D-printed substrates adapt to variations in infill density as shown in Figure 2.11. Increasing the infill percentage from 20% to 60% causes a general decrease in coating height Figure 2.11). Approximately 37% of particles are deformed at a 20% infill density, getting a diameter more than average diameter of feedstock particles, and approximately 65% of particles are highly deformed at a 60% infill density, both of which are consistent with the decrease in coating height as seen in Figure 2.12. The infill density appears to be the deciding factor in how much the particles deform as panel stiffness increases. The presented substrate distortion measurement results are evidence of this. In fact, panels with a 20% infill density exhibit higher distortion values than

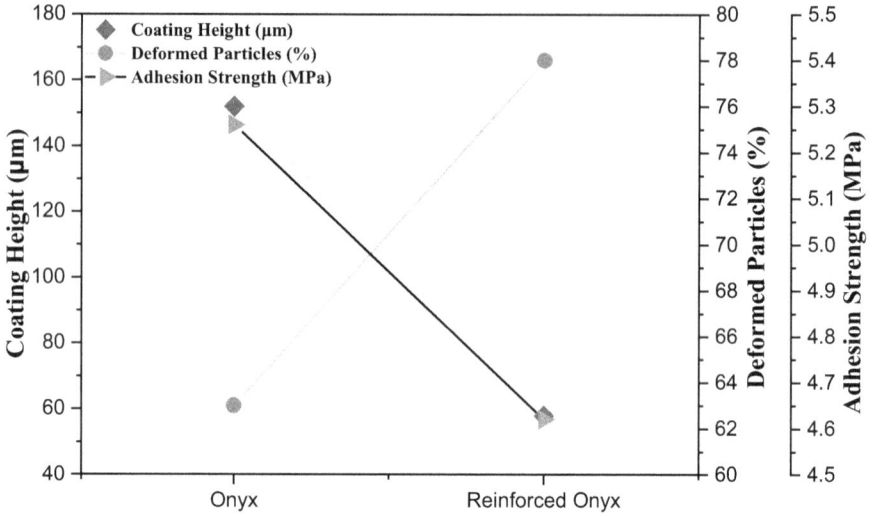

FIGURE 2.10   Impact of the Cold Spray deposition on fibres.

panels with 40% and 60% infill densities, suggesting an increase in panel stiffness with an increase in infill density. This holds true regardless of the thickness of
the Onyx layer. Despite the fact that the results in the preceding section show that
stronger adhesion is produced on the less rigid panels with the particles that should
profoundly penetrate the substrate, the adhesion values increase with increasing

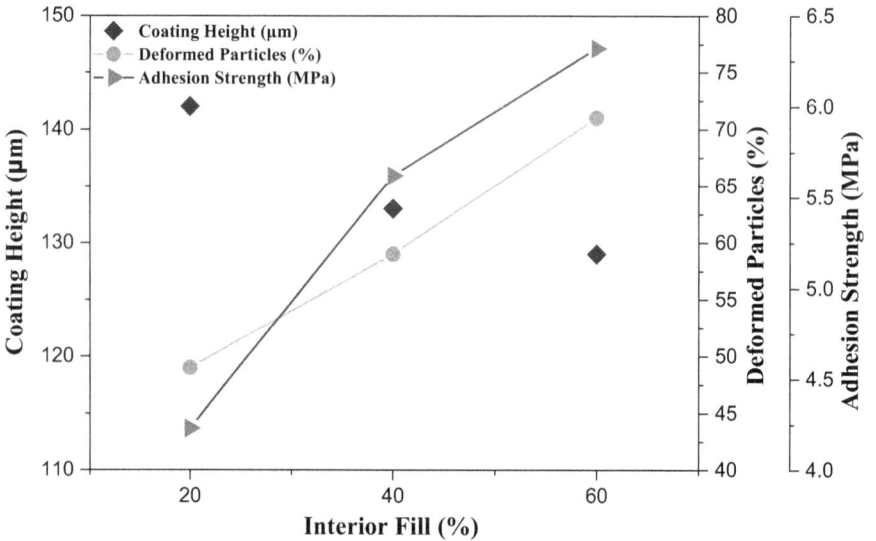

FIGURE 2.11   Effect of interior fill percentage on various factors.

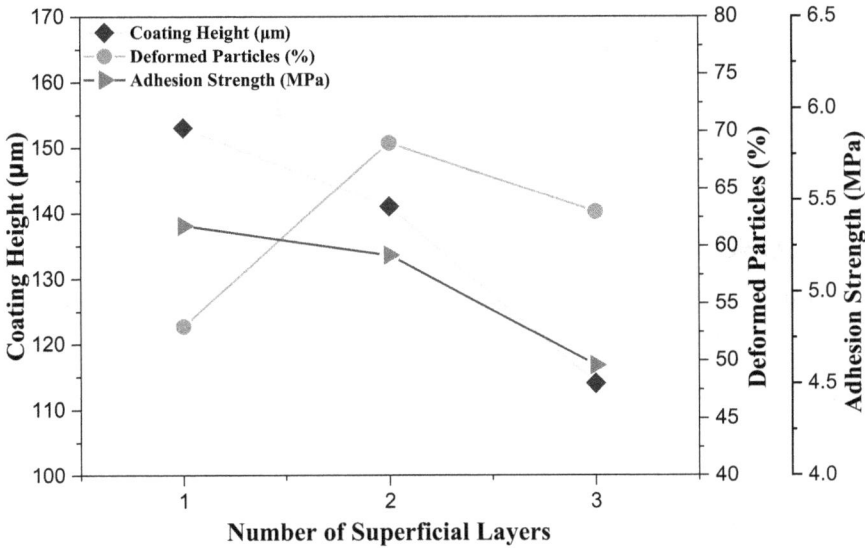

FIGURE 2.12   Impact of superficial matrix layer thickness.

infill density. It appears from the low distortion during deposition that the stiffening impact of the fibres dominates over the stiffening impact of increasing the interior fill density. The adhesion values are not just determined by the stiffening effect; there is another phenomenon at play. According to the literature reviews (Eren et al. 2023; Ansari & Kamil 2023), the adhesion mechanisms can be significantly affected by the surface texturing. Particles become entangled between two consecutive polymer tracks beginning at an infill density of 20% and increasing to 60%.

Since the entangled particles are more challenging to remove from the denser panels, we can infer that the coating adheres better to them. This research found that bonding between metal particles and polymer substrate could be improved, leading to a raise in bonding strength up to 40%, by reducing distance between polymer tracks.

2.4.3   EFFECT OF MATRIX LAYER THICKNESS

Figure 2.12 shows the outcomes of Cold Spray deposition of polymers to metals with varying thicknesses of the superficial matrix layer. When two layers of Onyx or an extra layer of matrix are deposited on the panel's top surface, the coating height is reduced. Particles experience significant distortion as the number of surface layers increases, as shown in Figure 2.12. Deformation of the particles is clearly visible after deposition of just one layer on a panel, with about 45% affected, and after deposition of two layers, with 52% affected. After the third coat of Onyx is sprayed, the coating seems severely deformed, with 50% of splats having a larger diameter than the average diameter of the process powders. The normal behaviour for coat thickness is consistent with this pattern. Substrate deformation was reduced in the three-layer panel, indicating that the surface matrix's thickness is crucial to the substrate's stiffening. Adding another layer to the surface has a similarly negative effect on adhesion. The stiffness of the panels is to blame for this, and it tends to increase in relation to the number of their outer layers. Stronger adhesion values are observed when particles strike a softer substrate and are more likely to penetrate the surface of polymer and anchor with neighbouring polymer materials.

2.5 CONCLUSIONS

Scientists looked into the effects of PMC production methods on the metallization of three-dimensional-(3D-)printed panels by using deposition of Cold Spray on titanium (Ti) particles. Focus was placed on the role played by the polymer's reinforcing fibres, as well as the infill density and top layer thickness. The following conclusions can be drawn from the aforementioned experimental results:

- It has been demonstrated that, under ideal processing conditions, Fused Filament Fabrication (FFF) of GFRP can be combined with CSD of titanium particles to achieve surface cover values nearer to 100% and bonding strength values near to 4 MPa.
- The composite substrate lay-up sequence plays a crucial role in fabrication of hybrid structures, with sequence being more crucial than deposition conditions.
- For instance, the bonding strength between metal particles and polymer substrate increases by 40% when the infill density is raised from 20% to 60%.
- In order to create more uniform metallic coatings, it was necessary to increase thickness of superficial matrix layer, which in turn reduces the adhesion to the polymeric substrate. Adhesion strength can decrease by as much as 25% when expanding from a single superficial layer to three.
- Since the fibres make the substrate more rigid, distortion phenomena are prevented while the material is being deposited.
- However, problems with coating growth mean that the maximum thickness that can be achieved is > 100 µm.
- Strengthening the bond between the powders and the substrate is essential if these structures are to find more widespread use.
- Coating growth must also be investigated, as only thin coatings have been achieved thus far.
- Thermoset polymers, on which deposition can also be performed, merit study.

REFERENCES

Ahmadifar, M., Benfriha, K., Shirinbayan, M., Fitoussi, J., & Tcharkhtchi, A. (2022). Mechanical behavior of polymer-based composites using fused filament fabrication under monotonic and fatigue loadings. *Polymers and Polymer Composites*, *30*, 1–11.

Ansari, A. A., & Kamil, M. (2023). Dimensional aspect of feedstock material filaments for FDM 3D printing of continuous fiber-reinforced polymer composites. In *3D Printing* (pp. 21–34). CRC Press.

Araya-Calvo, M., López-Gómez, I., Chamberlain-Simon, N., León-Salazar, J. L., Guillén-Girón, T., Corrales-Cordero, J. S., & Sánchez-Brenes, O. (2018). Evaluation of compressive and flexural properties of continuous fiber fabrication additive manufacturing technology. *Additive Manufacturing*, 22, 157–164.

Borg, G., Kiss, S., & Rochman, A. (2021). Filament development for laser assisted FFF 3D printing. *Journal of Manufacturing and Materials Processing*, 5(4), 115.

Dharmalingam, G., Prasad, M. A., Deepak, P., Sachin, S., Siva kumar, M., & Rathinasuriyan, C. (2023). Optimization of AWJM process parameters on 3D-printed onyx-glass fiber hybrid composite. *Journal of Harbin Institute of Technology (New Series)*, 30(2), 84–98.

Eren, Z., Burnett, C. A., Wright, D., & Kazancı, Z. (2023). Compressive characterisation of 3D printed composite materials using continuous fibre fabrication. *International Journal of Lightweight Materials and Manufacture*, 6(4), 494–507.

Grujovic, N., Zivic, F., Zivkovic, M., Sljivic, M., Radovanovic, A., Bukvic, L., Mladenovic, M., & Sindjelic, A. (2017). Custom design of furniture elements by fused filament fabrication. *Proceedings of the Institution of Mechanical Engineers, Part C: Journal of Mechanical Engineering Science*, 231(1), 88–95.

Horňáček, L., Halama, R., Krzikalla, D., Měsíček, J., Pauldoss, J., & James, N. (2022). Validation of methodology for numerical modeling of 3D printed structures reinforced by long carbon fibers. In *EAN 2021 - 59th International Scientific Conference on Experimental Stress Analysis - Book of Full Papers* (pp. 58–63).

Ichihara, N., & Ueda, M. (2022). 3D-print path generation of curvilinear fiber-reinforced polymers based on biological pattern forming. In *ECCM 2022 - Proceedings of the 20th European Conference on Composite Materials: Composites Meet Sustainability*, 1:401–407.

Imaeda, Y., Todoroki, A., Matsuzaki, R., Ueda, M., & Hirano, Y. (2021). Modified moving particle semi-implicit method for 3D print process simulations of short carbon fiber/polyamide-6 composites. *Composites Part C: Open Access*, 6, 1–10.

Jia, Z., Wang, Q., & Liu, J. (2023). An investigation of printing parameters of independent extrusion type 3D print continuous carbon fiber-reinforced PLA. *Applied Sciences*, 13(7), 4222.

Joch, R., Šajgalík, M., Drbúl, M., Holubják, J., Czán, A., Bechný, V., & Matúš, M. (2023). The application of additive composites technologies for clamping and manipulation devices in the production process. *Materials*, 16(10), 3624.

Kalman, H., Tripathi, N. M., Gabrieli, O. G., & Portnikov, D. (2017). Phase diagrams for pneumatic and hydraulic conveying. In *18th International Conferences on Transport and Sedimentation of Solid Particles, T and S 2017* (pp. 145–152). Wydawnictwo Uniwersytetu Przyrodniczego we Wroclawiu.

Kang, S., Cha, H., Ryu, S., Kim, K., Jeon, S., Lee, J., & Kim, S. (2022). Evaluation of cryogenic tensile properties of composite materials fabricated by fused deposition modeling 3D printer. *Progress in Superconductivity and Cryogenics (PSAC)*, 24(1), 8–12.

Kohn, O., Rosenthal, Y., Ashkenazi, D., Shneck, R., & Stern, A. (2021). Fused filament fabrication additive manufacturing: Mechanical response of polyethylene terephthalate glycol. *Annals of "Dunarea de Jos" University of Galati. Fascicle XII, Welding Equipment and Technology*, 32, 47–55.

Kumar, M. H., Mathivanan, N. R., & Kumar, S. (2018). Experiment investigations of effect of laminate thickness on flexural properties of GLARE and GFRP laminates. *Materials Research Express*, 6(2), 025313.

Lee, G. W., Kim, T. H., Yun, J. H., Kim, N. J., Ahn, K. H., & Kang, M. S. (2023). Strength of onyx-based composite 3D printing materials according to fiber reinforcement. *Frontiers in Materials, 10*, 1183816.

Li, J. H., Durandet, Y., Huang, X. D., & Ruan, D. (2022). Tensile properties of additively manufactured continuous glass fiber reinforced onyx. In *Materials Science Forum* (Vol. 1054, pp. 15–20). Trans Tech Publications Ltd.

Lumay, G., Tripathi, N. M., & Francqui, F. (2019). How to gain a full understanding of powder flow properties, and the benefits of doing so. *ONdrugDelivery, 2019*, 42–47.

Manikandan, R., Ponnusamy, P., Nanthakumar, S., Gowrishankar, A., Balambica, V., Girimurugan, R., & Mayakannan, S. (2023). Optimization and experimental investigation on AA6082/WC metal matrix composites by abrasive flow machining process. *Materials Today: Proceedings*.

Miri, S., Kalman, J., Canart, J. P., Spangler, J., & Fayazbakhsh, K. (2022). Tensile and thermal properties of low-melt poly aryl ether ketone reinforced with continuous carbon fiber manufactured by robotic 3D printing. *The International Journal of Advanced Manufacturing Technology, 122*(2), 1041–1053.

M Tripathi, N., & S Mallick, S. (2017). Pneumatic conveying of Fly Ash: Bend Models investigation. *Advanced Materials Proceedings, 2*(8), 526–531.

Muruganandhan, P., Jothilakshmi, S., Vivek, R., Nanthakumar, S., Sakthi, S., Mayakannan, S., & Girimurugan, R. (2023). Investigation on silane modification and interfacial UV aging of flax fibre reinforced with polystyrene composite. *Materials Today: Proceedings*.

Naik, M., Thakur, D., & Salunkhe, S. (2023). Evaluation of thermal and mechanical properties of continuous fiberglass reinforced thermoplastic composite fabricated by fused deposition modeling. *Journal of Applied Polymer Science, 140*(26).

Neveu, A., Lumay, G., Pillitteri, S., Monsuur, F., Pauly, T., Ribeyre, Q., Francqui, F., Vandewalle, N., & Tripathi, N. M. (2020b). Physical characterization of blends containing mesoporous particles with a focus on electrostatic properties. In *2020 AIChE Spring Meeting and 16th Global Congress on Process Safety*. AIChE.

Neveu, A., Tripathi, N. M., Rigo, O., Francqui, F., & Lumay, G. (2020a). Experimental investigation of spreadability of metal powders in recoating process. In *Proceedings - Euro PM2020 Congress and Exhibition (Proceedings - Euro PM2020 Congress and Exhibition)*. European Powder Metallurgy Association (EPMA).

Pace, F., Stamopoulos, A. G., Eckl, M., Senck, S., & Glinz, J. (2023). Analysis of the manufacturing porosity in fused filament fabricated onyx/long fiber reinforced composites using X-ray computed tomography. *Journal of Nondestructive Evaluation, 42*(4), 86.

Papa, I., Manco, E., Epasto, G., Lopresto, V., & Squillace, A. (2022). Impact behaviour and non destructive evaluation of 3D printed reinforced composites. *Composite Structures, 281*, 1–13.

Papa, I., Manco, E., Lopresto, V., Cigliano, C., Manzo, A., Silvestri, A. T., & Squillace, A. (2021, September). Penetration impact behaviour of innovative 3d printing onyx/glass composite samples. In *2021 IEEE 6th International Forum on Research and Technology for Society and Industry (RTSI)* (pp. 41–46). IEEE.

Pernica, J., Gajdaczek, T., Černý, P., Dostal, P., Brabec, M., & Korenko, M. (2023). Use of digital image correlation in predicting mechanical properties of parts made by additive manufacturing. *Acta Technologica Agriculturae, 26*(3), 166–172.

Prajapati, A. R., Dave, H. K., & Raval, H. K. (2020). Influence of fiber rings on impact strengthof 3d printed fiber reinforced polymer composite. *Manuf. Technol, 12*, 157–163.

Prajapati, A. R., Dave, H. K., & Raval, H. K. (2021). Effect of fiber reinforcement on the open hole tensile strength of 3D printed composites. *Materials Today: Proceedings, 46*, 8629–8633.

Prajapati, A. R., Dave, H. K., & Raval, H. K. (2023). Impact energy absorption and fracture mechanism of FFF made fiberglass reinforced polymer composites. *Rapid Prototyping Journal*, *29*(2), 275–287.

Radojičić, S., Konjatić, P., Katinić, M., & Kačmarčik, J. (2023). The influence of material storage on mechanical properties and deterioration of composite materials. *Tehnički vjesnik*, *30*(5), 1645–1651.

Ramesh, B., Kumar, S. S., Elsheikh, A. H., Mayakannan, S., Sivakumar, K., & Duraithilagar, S. (2022). Optimization and experimental analysis of drilling process parameters in radial drilling machine for glass fiber/nano granite particle reinforced epoxy composites. *Materials Today: Proceedings*, *62*, 835–840. https://doi.org/10.1016/j.matpr.2022.04.042

Rao, G. S., Paul, R., Singh, S., & Debnath, K. (2023). Influence of conventionally drilled and additively fabricated hole on tensile properties of 3D-printed ONYX/CGF composites. *Journal of Materials Engineering and Performance*, *32*(13), 5849–5861.

Rashid, A. A., & Koç, M. (2021). Creep and recovery behavior of continuous fiber-reinforced 3DP composites. *Polymers*, *13*(10), 1644.

Ratsimba, A., Zerrouki, A., Tessier-Doyen, N., Nait-Ali, B., André, D., Duport, P., Neveu, A., Tripathi, N., Francqui, F., & Delazir, G. (2021). Densification behaviour and three-dimensional printing of Y2O3 ceramic powder by selective laser sintering. *Ceramics International*, *47*(6), 7465–7474. https://doi.org/10.1016/j.ceramint.2020.11.087

Sharma, A., Babbar, A., Tian, Y., Pathri, B. P., Gupta, M., & Singh, R. (2022). Machining of ceramic materials: A state of the art review. *International Journal on Interactive Design and Manufacturing (IJIDeM)*, 1–21. https://doi.org/10.1007/s12008-022-01016-7

Sharma, A., & Jain, V. (2020). Experimental investigation of cutting temperature during drilling of float glass specimen. In *IOP Conference Series: Materials Science and Engineering* (Vol. 715, No. 1, p. 012050). IOP Publishing.

Sharma, A., Jain, V., & Gupta, D. (2018). Characterization of chipping and tool wear during drilling of float glass using rotary ultrasonic machining. *Measurement*, *128*. 254–263.

Sharma, A., Jain, V., & Gupta, D. (2019a). Tool wear analysis while creating blind holes on float glass using conventional drilling: A multi-shaped tools study. In *Advances in Manufacturing Processes: Select Proceedings of ICEMMM 2018* (pp. 175–183). Springer Singapore.

Sharma, A., Jain, V., & Gupta, D. (2019b). Comparative analysis of chipping mechanics of float glass during rotary ultrasonic drilling and conventional drilling: For multi-shaped tools. *Machining Science and Technology*, *23*(4), 547–568.

Sharma, A., Jain, V., & Gupta, D. (2019c). Multi-shaped tool wear study during rotary ultrasonic drilling and conventional drilling for amorphous solid. *Proceedings of the Institution of Mechanical Engineers, Part E: Journal of Process Mechanical Engineering*, *233*(3), 551–560.

Sharma, A., Jain, V., & Gupta, D. (2021a). Effect of pre and post tempering on hole quality of float glass specimen: For rotary ultrasonic and conventional drilling. *Silicon*, *13*, 2029–2039.

Sharma, A., Jain, V., & Gupta, D. (2021b). Mathematical approach on chipping volume estimation generated during rotary ultrasonic drilling for float glass. *Proceedings of the National Academy of Sciences, India Section A: Physical Sciences*, *92*, 285–291.

Sharma, A., Jain, V., Gupta, D., & Babbar, A. (2020). A review study on miniaturization. In *Advanced Manufacturing and Processing Technology* (First edition, pp. 111–131). CRC Press.

Sitotaw, D. B., Muenks, D., Kyosev, Y., & Kabish, A. K. (2022). Influence of fluorocarbon treatment on the adhesion of material extrusion 3D prints on textile. *Journal of Industrial Textiles*, *52*.

Tripathi, N. M., Francqui, F., & Lumay, G. (2020). Influence of relative air humidity on the flow property of fine powders. In *Third International Conference on Powder, Granule and Bulk Solids: Innovations and Applications PGBSIA 2020 February 26–28, 2020* (p. 63).

Tripathi, N. M., Francqui, F., Pirenne, T., & Lumay, G. (2019). Measuring Food Powders Electrical Properties as a result of anti-static content. In *2019 AIChE Annual Meeting*. American Institute of Chemical Engineers.

Tripathi, N. M., Levy, A., & Kalman, H. (2016). Initial acceleration pressure drop in dilute phase pneumatic conveying system. In *Powder, Granule and Bulk Solids: Innovations and Applications Conference*.

Tripathi, N. M., & Mallick, S. S. (2014). *An Investigation into Pressure Drop Across Bends for Fluidised Densephase Pneumatic Conveying Systems* (Doctoral dissertation).

Vaško, M., Sága, M., Majko, J., Vaško, A., & Handrik, M. (2020). Impact toughness of FRTP composites produced by 3D printing. *Materials*, *13*(24), 5654.

3 Advanced Characteristics of Materials Used in Additive Manufacturing
A State-of-the-Art Review and Future Applications

Neetesh Soni, Gilda Renna, and Paola Leo

3.1 INTRODUCTION

Additive Manufacturing (AM), also known as three-dimensional (3D) printing, is a revolutionary technology that has transformed the way products are designed and manufactured. With the ability to create complex shapes and structures layer by layer, Additive Manufacturing has become an eye-catching choice in a range of industries, including aerospace, automotive, and medical (Huang et al., 2015). However, the quality and performance of the final product largely depend on the materials used in the Additive Manufacturing process. As such, there is a growing need for advanced materials with specific characteristics to meet the demands of various applications. This chapter aims to provide an overview of the advanced characteristics of materials used in Additive Manufacturing, including their properties, processing techniques, and current applications. Furthermore, in addition to the advanced characteristics of these materials, an overview of the challenges faced in the development and application of advanced AM materials and potential solutions to overcome these challenges are also provided. Some of the advanced characteristics of materials that will be covered in this chapter include high strength and durability, heat resistance, electrical conductivity, biocompatibility, and shape memory (Blakey et al., 2021). These properties are essential for a wide range of applications, from lightweight and strong aerospace components to medical implants and devices. By understanding the characteristics and capabilities of these advanced materials, researchers, engineers, and designers can create more complex and functional products that meet the evolving needs of the modern world.

One of the most significant benefits of Additive Manufacturing is the ability to use a wide range of materials, including metals, plastics, ceramics, and composites, to create complex parts and components with high precision and accuracy. Various

DOI: 10.1201/9781003428862-3

materials used in Additive Manufacturing have unique properties that make them ideal for specific applications. For example, high-strength and durable materials like titanium and aluminium alloys are commonly used in the aerospace industry to create lightweight components that can withstand extreme temperatures and harsh environments (Bahnini et al., 2018). Similarly, materials with high heat resistance, such as ceramic composites and Carbon Fiber-Reinforced Polymers, are used in industries where the components need to operate at high temperatures, such as in engines, turbines, and exhaust systems. Electrical conductivity is another critical property that is required in electronic components, and materials like copper, silver, and gold are commonly used in Additive Manufacturing to create conductive parts and devices (Al-Furjan et al., 2022). ISome studies have addressed the advanced characteristics of materials used in pre- and post-processing during Additive Manufacturing (Sharma et al., 2022; Sharma & Jain, 2020; Sharma et al., 2018; Sharma et al., 2019a; Sharma et al., 2019b; Sharma et al., 2019c; Sharma et al., 2021a; Sharma et al., 2021b; Sharma et al., 2020; Sharma et al., 2022).

Another important characteristic of advanced materials in Additive Manufacturing is biocompatibility, which refers to the ability of a material to interact with biological systems without causing harm or adverse reactions. Biocompatible materials like titanium, stainless steel, magnesium alloys, and biodegradable polymers are used in the medical industry to create implants, prosthetics, and surgical instruments. Shape memory alloys, which can return to their original shape after being deformed, are also commonly used in medical devices like stents and orthodontic wires. Despite the advantages of using advanced materials in Additive Manufacturing, there are also challenges and limitations to be considered. For example, some materials may be difficult to process, requiring specialized equipment and techniques. Others may have limited availability or high costs, making them less feasible for large-scale production. The use of advanced materials in Additive Manufacturing has enabled the creation of complex and functional products that meet the evolving demands of various industries. Continuing to develop can unlock new possibilities for Additive Manufacturing, from personalized medical devices to high-performance aerospace components (Shah, et al., 2019).

Their current and potential future applications highlight the importance of selecting the appropriate material for a specific application, ensuring the desired functionality and performance of the final product. Moreover, the knowledge of the limitations associated with the use of advanced materials in Additive Manufacturing is of considerable help to direct future research and development efforts.

3.2   ROLE OF ADDITIVE MANUFACTURING: CLASSIFICATION

As the name implies, AM is the process of creating three-dimensional objects by adding layers of material one on top of the other. AM has been drawing much attention in the last two decades due to its cost-effectiveness, time, quality, production, and efficiency of the product. AM techniques can be classified into seven main categories, based on the type of material and the process used to create the object. There are several types of the Additive Manufacturing process, such as material extrusion, material jetting, binder jetting, powder bed fusion, sheet lamination, direct energy deposition, and hybrid manufacturing (i.e. Hybrid Laser Deposition and Machining (HLDM) and Laser Engineered Net Shaping (LENS)). The latter process is very promising, especially in the industrial field. Specifically, this combines metal AM technology with conventional subtractive technology, allowing each process to work together on the same part. Figure 3.1 shows a summary diagram of each AM process. Each of these techniques has a series of characteristics that make it different from the others, the most important of which may vary: the materials used for the printed component, the materials used for the supports, the printing speed, tolerances and surface roughness of the printed component, maximum print volume, ability to make final parts or prototypes only, and temperature variation. The latter being in turn a function of the material used during printing which affects the generation of heat and the continuity of production during the AM process (i.e. 3D printing for metal and non-metal AM process has different melting temperatures) (Zhang et al., 2018).

The classification of AM techniques provides a framework for selecting the most appropriate method for a particular application, taking into account factors such as material properties, production volume, and cost. The different methods of AM have advantages and disadvantages, each with respect to the other, attributable to the specific characteristics of the production process; however, by contrasting them with traditional techniques, it is possible to identify a series of benefits and limitations attributable to the entire category.

In particular, there are several advantages to using 3D printing such as:

1. Near-net-shape accuracy of the objects can be achieved.
2. Pre-analysis of the 3D-printed objects' mechanical properties can be done by simulations.
3. This can create imagination in physical objects.
4. Time and cost are also important factors.

The detailed explanation and particular AM process are given below.

3.2.1 Material Jetting

Material Jetting is an AM technique that uses inkjet printing technology to deposit droplets of material onto a substrate layer by layer to create a 3D object. The material is usually a thermoplastic, resin, or wax that is heated and deposited through a print head onto a build platform. The print head moves back and forth across the platform, depositing the material in precise locations to build up the object.

Material Jetting is a highly precise and accurate AM method that allows for the creation of complex geometries and high-resolution parts with smooth surface finishes. The method can produce parts with varying colours, transparency, and rigidity. Material Jetting is used in various applications, including prototyping, tooling, and small-scale production of products, such as consumer goods, medical devices, and aerospace components (Gulcan & Aykut, 2021).

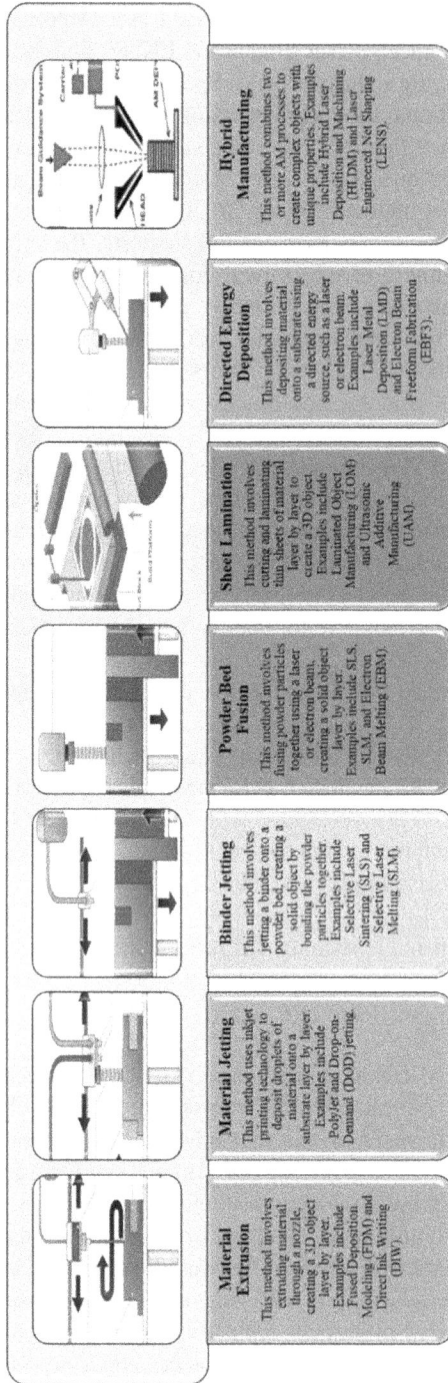

FIGURE 3.1   Classification of the different types of Additive Manufacturing process.

Merits of Material Jetting:

1. High precision and accuracy.
2. Capable of producing parts with varying colours, transparency, and rigidity.
3. Can create complex geometries with smooth surface finishes.
4. Suitable for small-scale production and prototyping.
5. Can print multiple materials at once, allowing for the creation of multi-material parts.

Limitations of Material Jetting:

1. Limited build volume compared to other AM techniques.
2. Limited material selection compared to other AM techniques.
3. Parts may require post-processing to remove support structures and excess material.
4. Can be expensive for high-volume production.

Overall, Material Jetting is a useful AM method for producing high-precision and complex parts with smooth surface finishes. While it has some limitations, its ability to print multiple materials at once and create multi-material parts makes it a useful tool in the manufacturing industry. Its applications range from prototyping to small-scale production, with the potential for further advancements in the future.

3.2.2 POWDER BED FUSION

Powder Bed Fusion (PBF) uses a laser or electron beam to selectively fuse layers of metal, plastic, or ceramic powder together, layer by layer, to create a 3D object. In PBF, a thin layer of powder is spread over a build platform, and a laser or electron beam is used to melt and fuse the powder particles together based on a 3D digital model. The build platform is then lowered, and a new layer of powder is spread, and the process is repeated until the object is complete.

Powder Bed Fusion is a versatile AM technique that is widely used in various applications, including aerospace, medical, and automotive industries. It can produce parts with high complexity, intricate geometries, and excellent mechanical properties. The method is particularly useful for creating parts with fine features, thin walls, and intricate internal structures, such as lattice structures.

Merits of Powder Bed Fusion:

1. High accuracy and precision.
2. Excellent mechanical properties.
3. Capable of producing complex geometries and intricate internal structures.
4. Suitable for high-value, low-volume production.
5. Can use a wide range of materials, including metals, plastics, and ceramics.

Limitations of Powder Bed Fusion:

1. Limited build volume compared to other AM techniques.
2. Can be expensive for high-volume production.
3. Requires post-processing to remove support structures and excess powder.
4. Limited to certain materials and sizes.
5. Can be prone to warping or cracking during the printing process.

Overall, Powder Bed Fusion is an advanced AM technique with many applications in the manufacturing industry, particularly in the production of high-value, low-volume parts with intricate geometries. While it has some limitations, its ability to produce parts with excellent mechanical properties and complex geometries makes it a useful tool in the manufacturing industry. As the technology advances, it is likely to become more widely used in a range of industries, with increased speed and precision, and a wider range of materials.

3.2.3 SHEET LAMINATION

Sheet Lamination uses sheets of material, such as paper, plastic, or metal, that are bonded together layer by layer to create a 3D object. In Sheet Lamination, a layer of adhesive is applied to a sheet of material, and the sheet is then cut or shaped according to the 3D model. The process is repeated layer by layer until the object is complete.

Sheet Lamination is a cost-effective and efficient AM technique that is particularly useful for producing large, low-cost parts quickly. It can produce parts with relatively high strength, stiffness, and stability, and it is widely used in applications such as architecture, product design, and art (Gibson et al., 2010).

Merits of Sheet Lamination:

1. Cost-effective and efficient for producing large parts.
2. High level of precision and accuracy.
3. Wide range of materials can be used, including paper, plastic, and metal.
4. Suitable for low-cost, low-volume production.
5. Good strength, stiffness, and stability of the parts.

Limitations of Sheet Lamination:

1. Limited accuracy compared to other AM techniques.
2. Limited ability to produce intricate details or internal structures.
3. May require post-processing to achieve the desired surface finish.
4. Limited range of materials compared to other AM techniques.

Overall, Sheet Lamination is a useful AM technique for producing large, low-cost parts quickly, with a high level of precision and accuracy. While it has some limitations, its ability to produce parts with relatively high strength, stiffness, and stability makes it a useful tool in the manufacturing industry. Its applications range from architecture to art, with potential for further advancements in the future. As the technology advances, it is likely to become more widely used in a range of industries, with increased accuracy and precision, and a wider range of materials.

3.2.4 MATERIAL EXTRUSION

Material Extrusion uses a nozzle to extrude a melted material, such as plastic or metal, layer by layer, to create a 3D object. In Material Extrusion, a spool of filament is fed into a heated nozzle, where it is melted and extruded according to a 3D digital model. The nozzle moves along the X, Y, and Z axis to deposit the material layer by layer until the object is complete.

Material Extrusion is a versatile AM technique that is widely used in various applications, including prototyping, product design, and small-scale production. It can produce parts with relatively high accuracy and precision, and it is particularly useful for creating parts with simple geometries, such as housings or brackets.

Merits of Material Extrusion:

1. Low cost and high availability of materials.
2. Good accuracy and precision.
3. Easy to use and relatively simple to set up.
4. Suitable for low-cost, low-volume production.
5. Can produce parts with a range of mechanical properties.

Limitations of Material Extrusion:

1. Limited accuracy compared to other AM techniques.
2. Limited ability to produce intricate details or internal structures.
3. Limited range of materials compared to other AM techniques.
4. Can result in poor surface finish and texture.

Overall, Material Extrusion is a useful AM technique for producing low-cost, low-volume parts quickly and efficiently, with a reasonable level of accuracy and precision. While it has some limitations, its ability to produce parts with a range of mechanical properties and ease of use makes it a useful tool in the manufacturing industry. Its applications range from prototyping to small-scale production, with potential for further advancements in the future. As the technology advances, it is likely to become more widely used in a range of industries, with increased accuracy and precision, and a wider range of materials.

3.2.5 BINDER JETTING

Binder Jetting uses a liquid binding agent, such as glue or resin, to bond together layers of powder material, such as metal, ceramic, or polymer, to create a 3D object.

In Binder Jetting, a layer of powder material is spread over a build platform, and a liquid-binding agent is selectively deposited using a print head or inkjet printhead. The process is repeated layer by layer until the object is complete.

Binder Jetting is a versatile AM technique that is widely used in various applications, including prototyping, product design, and small-scale production. It can produce parts with relatively high accuracy and precision, and it is particularly useful for creating parts with complex geometries, such as lattice structures (Gokuldoss et al., 2017).

Merits of Binder Jetting:

1. Low cost and high availability of materials.
2. High speed of production.
3. Can produce complex geometries with relative ease.
4. Can use a wide range of materials, including metals, ceramics, and polymers.
5. Suitable for both low- and high-volume production.

Limitations of Binder Jetting:

1. Limited accuracy compared to other AM techniques.
2. Limited ability to produce intricate details or internal structures.
3. May require post-processing to achieve the desired surface finish.
4. Limited strength and durability of parts compared to other AM techniques.

Overall, Binder Jetting is a useful AM technique for producing complex, low-cost parts quickly and efficiently, with a reasonable level of accuracy and precision. While it has some limitations, its ability to produce parts with a wide range of materials and geometries makes it a useful tool in the manufacturing industry. Its applications range from prototyping to high-volume production, with potential for further advancements in the future. As the technology advances, it is likely to become more widely used in a range of industries, with increased accuracy and precision, and improved strength and durability of the parts.

3.2.6 DIRECT ENERGY DEPOSITION

Direct Energy Deposition (DED) involves the deposition of molten material, such as metal or plastic, onto a substrate to create a 3D object. In DED, a focused energy source, such as a laser or electron beam, is used to melt the material as it is deposited layer by layer.

DED is a versatile AM technique that is widely used in various applications, including aerospace, automotive, and medical industries. It can produce parts with a range of mechanical properties, including high strength and hardness, and it is particularly useful for creating large-scale parts with complex geometries.

Merits of Direct Energy Deposition:

1. High-speed production compared to other AM techniques.
2. Ability to produce large-scale parts with complex geometries.
3. Suitable for high-strength and high-hardness materials.
4. Can produce parts with a range of mechanical properties.
5. Can be used to repair existing parts.

Limitations of Direct Energy Deposition:

1. Limited accuracy compared to other AM techniques.
2. Limited resolution compared to other AM techniques.
3. Limited range of materials compared to other AM techniques.
4. May require post-processing to achieve the desired surface finish and texture.

Overall, Direct Energy Deposition is a useful AM technique for producing large-scale parts with complex geometries and high strength and hardness. While it has some limitations, its ability to produce parts quickly and efficiently makes it a useful tool in the manufacturing industry. Its applications range from aerospace and automotive to medical industries, with potential for further advancements in the future. As the technology advances, it is likely to become more widely used in a range of industries, with increased accuracy and precision, and a wider range of materials.

3.2.7 Vat Photo Polymerization

Vat Photo Polymerization involves using a photopolymer resin that is selectively cured by a light source, such as a laser or an UltraViolet (UV) light, to create a 3D object. In this process, a vat of liquid photopolymer resin is exposed to a pattern of light that selectively cures the material layer by layer, solidifying the resin and creating the object.

Vat Photo Polymerization is a versatile AM technique that is widely used in various applications, including jewellery, dental, and medical industries. It can produce parts with high resolution, accuracy, and detail, and it is particularly useful for creating parts with intricate geometries and high surface finish (Schwartz et al., 2022).

Merits of Vat Photo Polymerization:

1. High resolution and accuracy.
2. Can produce parts with intricate geometries and high surface finish.
3. Can produce parts with a range of mechanical properties.
4. Can be used to produce small-scale parts.

Limitations of Vat Photo Polymerization:

1. Limited range of materials compared to other AM techniques.
2. Limited build size compared to other AM techniques.

3. Slow production compared to other AM techniques.
4. Parts may be brittle and not suitable for high-stress applications.

Overall, Vat Photo Polymerization is a useful AM technique for producing high-resolution and highly detailed parts with intricate geometries and high surface finish. While it has some limitations, its ability to produce parts with a range of mechanical properties makes it a useful tool in the manufacturing industry. Its applications range from jewellery, dental, and medical industries, with potential for further advancements in the future. As the technology advances, it is likely to become more widely used in a range of industries, with increased accuracy and precision, and a wider range of materials.

3.3   TYPES OF MATERIALS INVOLVEMENT IN AM

There are several types of materials such as polymer, metals, ceramics, composites, and biomaterials that can be used in Additive Manufacturing processes. Listed below are some examples:

1. **Polymers**: Polymers are a type of material that is commonly used in AM processes due to its ability to be melted and extruded. Examples of polymers used in AM include Acrylonitrile-Butadiene-Styrene (ABS), PolyCarbonate (PC), and PolyAmide (PA).
2. **Metals**: Metals are also widely used in AM processes due to its ability to be melted and solidified. Examples of metal alloys used in AM include titanium alloys, aluminium alloys, and stainless steel.
3. **Ceramic**: Ceramic is a type of material that is becoming increasingly popular in AM processes due to its ability to withstand high temperatures and corrosive environments. Examples of ceramic used in AM include zirconia (ZrO2), alumina (Al2O3), and silicon nitride (Si3N4) (Ewais et al., 2010).
4. **Composites**: Composites are materials that consist of two or more different materials, such as fibres and a matrix material. Examples of composite materials used in AM include Carbon Fibre-Reinforced Polymers (CFRP) and Glass Fibre-Reinforced Polymers (GFRP) (Adeniran et al., 2021).
5. **Biomaterials**: Biomaterials are materials that are used in medical applications, such as implants and prosthetics. Examples of biomaterials used in AM include HydroxyApatite (HA), Polylactic Acid (PLA), and PolyGlycolic Acid (PGA).

These are just a few examples of the types of materials that can be used in the AM processes. The specific materials used will depend on the application, the desired properties of the final product, and efficiency of the AM technology being used. Figure 3.2 represents material types and the particular area of the application. The global demands, application, and achievement amount till 2021 is approx. $13.8 billion. This will be enhanced in the upcoming decades due to the high demand for lightweight structure and near-net shape components with higher precision accuracy

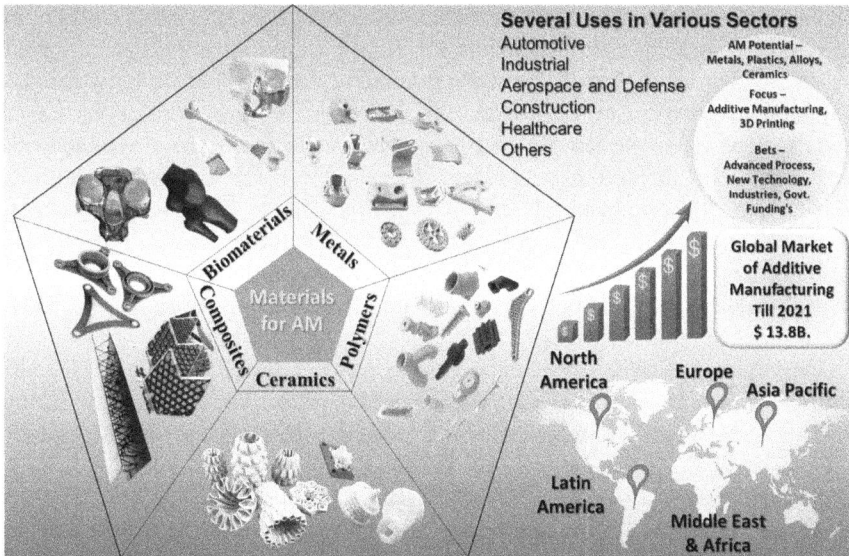

FIGURE 3.2 Additive Manufacturing materials application and worldwide demands.

of 3D-printed components. Day by day several areas are focusing on AM-3D-printed technology due to their merits such as time, cost, materials, operation etc. in details mentioned below about the particular materials and focusing AM process with proper justification of the applications and limitations. This technology is growing in the market and covers all the regions globally such as the US, Europe, Asia, Africa, Middle East, and Australia. Moreover, AM has many potentials to cover the market and several industries. Due to the freedom of materials utility, and single technology principle, it requires skills and promoting funding, and government supports (report analysis 2022–2030).

3.3.1 Metallic Materials for AM

Metallic materials are used in AM processes by melting the metal powder or wire and solidifying it layer by layer to form a 3D object. This process can be accomplished using a range of AM techniques, including Powder Bed Fusion, Directed Energy Deposition, and Binder Jetting. There are several applications of metallic materials, including in the aerospace, medical, and automotive industries. For example, titanium alloys are often used in aerospace due to their high strength-to-weight ratio and excellent corrosion resistance. Stainless steel is commonly used in medical implants due to its biocompatibility and corrosion resistance. Aluminium alloys from the 7xxx series are widely utilized as lightweight materials for manufacturing complex-shaped components through Additive Manufacturing (AM). Specifically, the AlSi10Mg alloy stands out as the most extensively employed option among the various available aluminium alloys. This preference is attributed to its remarkable strength, excellent stiffness-to-weight ratio, and commendable corrosion resistance. The main advantages of using metallic materials in AM include the ability to produce complex geometries, reduced material waste compared to traditional manufacturing processes, and the ability to create parts with unique properties such as lattice structures for lightweight applications. AM also enables the production of highly customized parts on demand. Moreover, there are some limitations to using metallic materials in AM which include the high cost of the materials, limited availability of metal powders, and difficulties in achieving consistent mechanical properties due to factors such as microstructural variability and residual stresses. Additionally, the high energy required to melt the metal can result in thermal distortion or cracking.

Also, the use of metallic materials in AM offers many advantages, such as the ability to produce complex geometries and unique properties. However, the cost of materials and issues related to mechanical properties and thermal distortion should be considered when selecting metallic materials for AM applications. As the technology continues to advance, it is likely that these limitations will be addressed, and metallic materials will become increasingly important in the AM industry (report analysis 2022–2030).

3.3.2 POLYMER MATERIALS FOR AM

The most common polymer materials used in AM include acrylonitrile-butadiene-styrene (ABS), polycarbonate (PC), and polyamide (PA). Polymer materials are used in AM processes by melting the polymer and extruding it layer by layer to form a 3D object. This process can be accomplished using a range of AM techniques, including Fused Deposition Modeling (FDM) and Material Jetting. Polymer materials are used in a variety of applications, including aerospace, automotive, medical, and consumer products. For example, PA is often used in the automotive industry due to its high strength and stiffness, while ABS is commonly used in consumer products such as toys and phone cases due to its low cost and ease of processing. The main advantages of using polymer materials in AM consist of the ability to produce complex geometries, reduced material waste compared to traditional manufacturing processes, and the ability to create parts with unique properties such as lattices for lightweight applications. Additionally, the relatively low cost of polymer materials makes them an attractive option for rapid prototyping. Some limitations of using polymer materials in AM include the limited range of mechanical properties compared to metals, limitations in achieving high-temperature resistance and dimensional stability, and potential issues with part accuracy and surface finish.

Furthermore, the use of polymer materials in AM offers many advantages, such as the ability to produce lightweight and complex geometries. However, the limited range of mechanical properties and potential issues with part accuracy and surface finish should be considered when selecting polymer materials for AM applications. As the technology continues to advance, it is likely that these limitations will be addressed, and polymer materials will become increasingly important in the AM industry (report analysis 2022–2030; Saleh et al., 2021).

3.3.3 CERAMIC MATERIALS FOR AM

The most common ceramic materials used in AM include zirconia, alumina, and silicon carbide. Ceramic materials are used in AM processes by melting or sintering the ceramic powder or filament to form a 3D object. This process can be accomplished using a range of AM techniques, including Powder Bed Fusion and Material Jetting. Ceramic materials are used in a variety of applications, including aerospace, medical, and electronics industries. For example, zirconia is often used in biomedical implants due to its biocompatibility and strength, while silicon carbide is used in electronic components due to its excellent thermal conductivity. The main advantages of using ceramic materials in AM include the ability to produce complex geometries with high strength and thermal resistance. Ceramic materials can also offer benefits such as corrosion resistance and biocompatibility, making them ideal for use in harsh environments or biomedical applications. Some limitations of using ceramic materials in AM include the high cost of materials and the limited availability of ceramic powders. Additionally, the brittleness of ceramic can make them difficult to work with and also the fact that they are prone to cracking during the printing process. The high temperatures required for processing ceramic can also lead to challenges with maintaining dimensional accuracy.

Overall, the use of ceramic materials in AM offers many advantages, such as high strength and thermal resistance, making them ideal for use in a variety of applications. However, the high cost of materials and potential issues with cracking and dimensional accuracy should be considered when selecting ceramic materials for AM applications. As the technology continues to advance, these limitations will be addressed, and ceramic materials will become increasingly important in the AM industry (Lakhdar et al., 2021).

3.3.4 COMPOSITE FOR AM

Composite for Additive Manufacturing (AM) refers to materials that combine two or more materials to create a hybrid material with unique properties. These materials can include combinations of metals, polymers, and ceramic. Composite for AM is used by combining multiple materials in a powder or filament form and depositing them layer by layer to form a 3D object. This process can be accomplished using a range of AM techniques, including Powder Bed Fusion and Material Jetting. Composite materials have a wide range of applications, including aerospace, automotive, and biomedical industries and electronics. For example, composite materials that combine carbon fibres and thermoplastic resins can be used in lightweight and high-strength parts for aircraft, while metal-polymer composites can be used in orthopaedic implants. The main advantage of using composite materials is the ability to create parts with unique and tailored properties. By combining materials, it is possible to create parts with specific mechanical, thermal, and electrical properties that cannot be achieved with single materials. Additionally, these materials can offer benefits such as reduced weight and improved biocompatibility. Some limitations of using composite and other materials for AM include the difficulty in achieving consistent material properties throughout the printed part, as well as the potential for delamination or separation of the different materials during the printing process. Additionally, the cost of composite materials can be higher than single-material options.

Additionally, composite, and other materials for AM offer the ability to create parts with unique and tailored properties. While there are limitations and potential challenges in working with these materials, their benefits make them a valuable option in a range of industries. As AM technology continues to advance, it is likely that the use of composite and other materials will become increasingly important and prevalent.

3.3.5   3D Bioprinting Organ and Cells

3D bioprinting of organs and cells is a rapidly evolving field that holds great promise for the future of medicine. The technology involves printing layers of living cells, along with biomaterials, to create tissues and even entire organs. 3D bioprinting involves the use of a bioprinter to deposit living cells onto a scaffold, which can then be used to grow tissue or even whole organs. The process involves creating a digital model of the desired tissue or organ, which is then translated into a physical object using a bioprinter. 3D bioprinting has many potential applications in the field of medicine, including the production of tissues and organs for transplantation, drug testing and discovery, and disease modelling. For example, bio-printed skin tissue can be used to test the effectiveness and safety of cosmetic products, while bio-printed liver tissue can be used for drug toxicity testing. The main benefit of 3D bioprinting is the ability to create complex tissues and organs with precise control over their structure and function. This technology holds the potential to revolutionize organ transplantation by providing a solution to the shortage of donor organs. Additionally, 3D bioprinting can be used for drug discovery and personalized medicine, as bio-printed tissues and organs can be used to test the effectiveness of drugs and treatments on specific patients. Some boundaries of 3D bioprinting include the need for highly specialized equipment and expertise, as well as the challenges of creating functional blood vessels and nerves within bio-printed tissues and organs. Additionally, the cost of the materials and equipment required for bioprinting can be prohibitive.

Also, 3D bioprinting of organs and cells has the potential to transform the field of medicine by enabling the production of customized tissues and organs for transplantation, as well as providing a powerful tool for drug discovery and disease modelling. While there are still limitations and challenges that need to be overcome, the rapid development of this technology is opening up new possibilities for the future of medicine (Vyas & Udayawar, 2019).

3.3.6   Self-healing and Shape Memory for Artificial Organs

Self-healing and shape memory are two emerging technologies that hold great promise for the development of artificial organs. Self-healing materials have the ability to repair themselves when damaged, while shape memory materials can change their shape in response to external stimuli, such as temperature or pH. These technologies can be incorporated into artificial organs to enhance their functionality and durability. Self-healing and shape memory technologies can be used in a variety of artificial organs, including heart valves, blood vessels, and even entire organs such as the liver and kidneys. These technologies can help to extend the lifespan of artificial organs, reduce the need for replacement surgeries, and improve overall patient outcomes. The advantage of self-healing and shape memory technologies in artificial organs is their ability to improve the lifespan and functionality of these devices. Self-healing materials can repair themselves when damaged, reducing the need for replacement surgeries, while shape memory materials can change their shape in response to changing physiological conditions, improving their compatibility with the human body. Restrictions on self-healing and shape memory technologies in artificial organs include the need for specialized materials and manufacturing processes, as well as the challenge of integrating these technologies into complex organ systems. Additionally, there may be concerns about the long-term safety and reliability of these materials.

Overall, self-healing and shape memory technologies hold great promise for the development of artificial organs that are more durable and compatible with the human body. While there are still challenges and limitations that need to be overcome, the rapid development of these technologies is opening up new possibilities for the future of organ transplantation and healthcare (Li et al., 2019).

3.4 FUTURISTIC ASPECT FOR ADDITIVE MANUFACTURING

Additive Manufacturing has already made significant advancements in the manufacturing industry, but there are several futuristic aspects that hold great promise for the future of this technology, for example, multi-material printing, 4D printing, Nanoscale printing, Artificial Intelligence (AI) and Machine Learning, In-situ characterization and monitoring, and bioprinting. Figure 3.3 explains the futuristic approach through the pillar diagram. The future development of additive manufacturing techniques is anticipated to bring about several advancements and enhancements in terms of characteristics. Here are some potential areas of progress:

1. *Increased Speed*: Efforts are being made to improve the printing speed of additive manufacturing processes.
2. *Larger Build Volumes*: Future developments aim to expand the build volumes of additive manufacturing machines, allowing for the creation of larger and more substantial parts in a single print.
3. *Improved Resolution*: Enhancements in resolution will enable the production of finer details and intricate geometries.
4. *Enhanced Material Selection*: Additive manufacturing techniques are continually expanding the range of materials that can be used. In the future, there will likely be an increase in the availability of specialized alloys,

Bioprinting

Bioprinting involves printing living cells and tissues to create functional organs or replace damaged tissues. The development of bioprinting technology holds great promise for the future of medicine and healthcare.

Nanoscale printing

The nanoscale level allowing for the creation of new materials and devices that were previously impossible to manufacture

4D printing

4D printing takes 3D printing to the next level by adding a fourth dimension of time. This means that objects printed using this technology can change their shape, size, or function in response to external stimuli such as temperature or humidity.

Artificial Intelligence (AI) and Machine Learning

AI and machine learning algorithms can optimize the design process and enhance the performance of AM techniques by predicting optimal printing conditions and improving the accuracy and precision of printed objects.

Multi-material printing

multi-material printing is an emerging technology that allows for the simultaneous printing of multiple materials which can result in more complex and functional

In-situ characterization and monitoring

In-situ characterization and monitoring technologies can provide real-time feedback on the printing process, enabling the identification and correction of errors before the object is completed.

Futuristic Approach

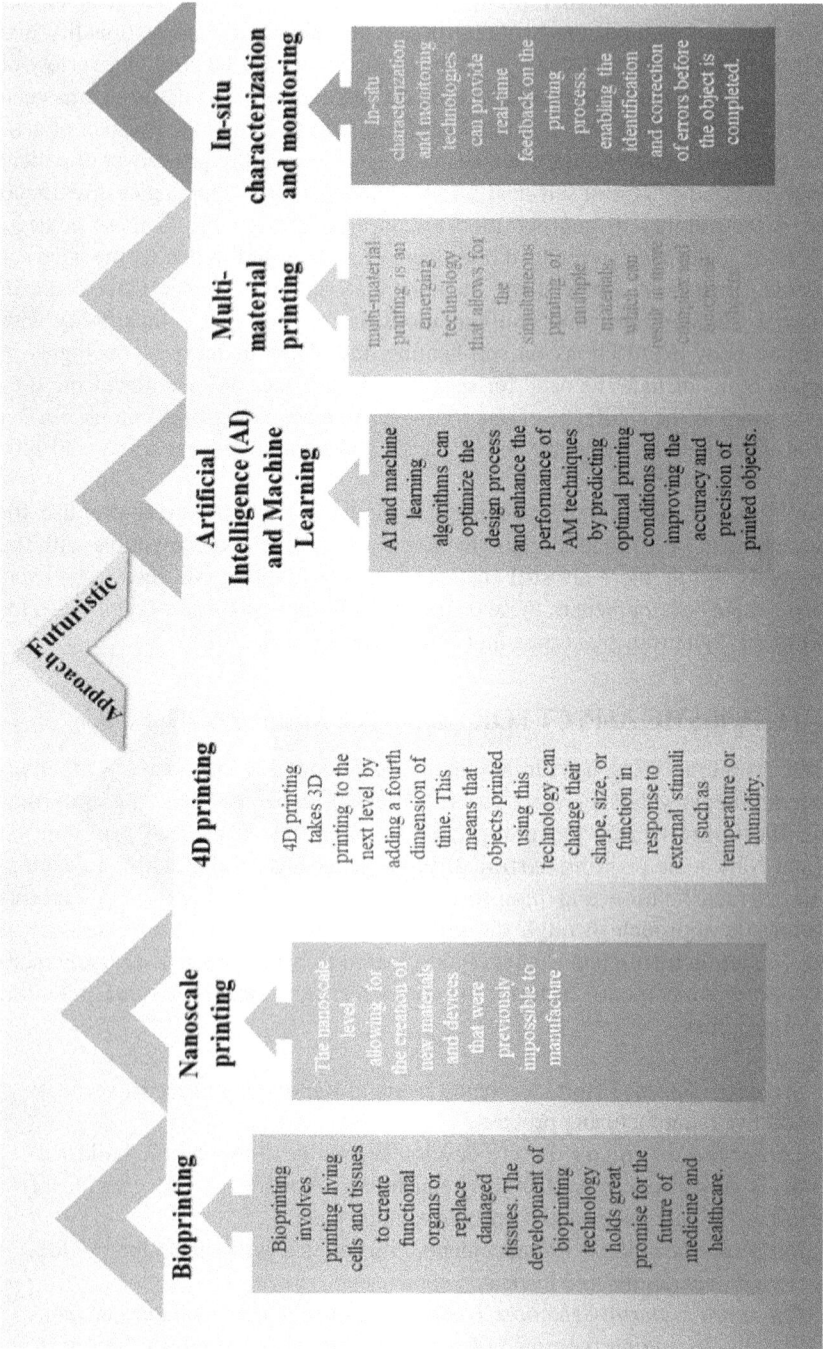

FIGURE 3.3 Futuristic approaches for the AM process.

composites, and advanced materials, further broadening the application possibilities.

5. *Integration of Post-Processing*: Additive manufacturing techniques are evolving to include post-processing steps within the same system. Integration of processes such as machining, surface finishing, and heat treatment would streamline the overall manufacturing workflow and reduce lead times.

6. *Sustainability and Recycling*: There is an increasing focus on developing more sustainable additive manufacturing practices, including the use of recycled materials, biodegradable polymers, and optimized energy consumption. Future advancements will likely prioritize environmentally friendly solutions.

Overall, the future of additive manufacturing holds promising prospects for faster production, improved resolution, expanded material selection, enhanced quality control, and increased sustainability, among other advancements.

3.5 CHARACTERISTICS OF THE AM-PRODUCED PARTS

3D printing or additive manufacturing is a process that creates a physical object from a digital model while using AM-based particular technology. There are several processes as mentioned earlier and characterizing of 3D-printed parts is important to ensure their quality, performance, and reliability. Different characterization techniques can be used to obtain specific information about the properties of the printed parts. A detailed characterization of AM-produced parts is mentioned in Table 3.1.

Overall, there are many different characterization techniques that can be used to evaluate 3D-printed parts, depending on the specific properties and requirements of the application. By using these techniques, designers and engineers can ensure that 3D-printed parts meet the desired specifications for their intended use and can continue to optimize the 3D-printing process for improved quality and performance. The important aspects of the AM have the potential to significantly reduce waste and improve the sustainability of the manufacturing process. By printing products on demand and using only the necessary number of materials, additive manufacturing can help to reduce the environmental impact of manufacturing. Also AM process has endless possibilities, and as the technology continues to evolve, we can expect to see many exciting and futuristic applications emerge in the coming years.

3.6 CONCLUSIONS

In conclusion, additive manufacturing has revolutionized the manufacturing industry, offering unique advantages such as the ability to create complex shapes and structures layer by layer. The quality and performance of the final product depend largely on the materials used in the process. This chapter has provided an overview of the characteristics of materials used in additive manufacturing, including their properties, processing techniques, current applications, and the challenges

TABLE 3.1

Important Characterization of AM-Produced Parts and Post Analysis

Characteristics	Objects	Process	Materials	Function	Remarks	Ref.
Dimensional accuracy	3D-printed dental casts	Material Jetting	Polyurethane dental cast	Clinical implications	This refers to how closely the printed part matches the intended dimensions. It can be measured using calipers or other precision measuring tools such as digital scanning camera and other visual scanners, computing methods	(Agarwal et al., 2016)
Surface quality	Engine intake flange	Fused filament fabrication process, extrusion-based AM	Polylactic acid (PLA)	Profilometer and redictivemodelling approach	The surface of a 3D-printed part may have small defects or imperfections that can affect its appearance or performance. Techniques like visual inspection, microscopy, or profilometry can be used to assess surface quality	(Bahnini et al., 2018)
Mechanical properties	Experimental cubic samples, masonry structures	Binder jet technology	Quartz-sand type GS19 and binderagent is a furan resin	AM helpful for the earthquake engineering prediction	The strength, stiffness, and toughness of a 3D-printed part can be determined through mechanical testing, such as tensile or compression testing	(Lakhdar et al., 2021)
Thermal properties	Cura profiles experimental samples	Fused Filament Fabrication (FFF)	Polylactic acid-metal composites– copper Fill, bronze Fill, stainless steel PLA, and magnetic iron PLA	Thermal conductivity analysis	Some 3D-printed materials may exhibit unique thermal properties, such as thermal conductivity or specific heat capacity. Thermal analysis techniques like Differential Scanning Calorimetry (DSC) or ThermoGravimetric Analysis (TGA) can be used to evaluate these properties	(Laureto et al., 2017)

(Continued)

TABLE 3.1 (CONTINUED)
Important Characterization of AM-Produced Parts and Post Analysis

Characteristics	Objects	Process	Materials	Function	Remarks	Ref.
Chemical properties	Experimental study	3D printing—PLA waste	Polylactic acid (PLA)	High calorific value with less impurities, distributedactivation energy model (DAEM) as well as Flynn-Wall and Ozawa (FWO), decomposed in temperature stages of 260–390 °C and 390–500 °C	3D-printed parts may be exposed to different chemicals or environments, depending on their intended use. Chemical analysis techniques like Fourier Transform Infrared Spectroscopy (FTIR) or Gas Chromatography-Mass Spectrometry (GC-MS) can be used to determine chemical composition and identify any degradation or changes in properties over time	(Bhardwaj et al., 2019)
Microstructural analysis	Permanent magnets specimens	Powder bed fusion technique	Isotropic Sm-Fe-N bonded magnets	Rare earths, investigated surface properties and analysis	3D-printed parts often have a unique microstructure due to the layer-by-layer additive manufacturing process. Microstructural analysis techniques like Scanning Electron Microscopy (SEM) or X-Ray Diffraction (XRD) can be used to observe and analyze the internal structure of the printed parts.	(Engerroff et al., 2019)
Electrical properties	Efficient energy devices, graphene-based devices	Extrusion-based 3D printing	Graphene-based electrodes, spherical copper powder	rCMG-based self-standingbinder-free electrode and coupled rCMG/Cu printed electrode prove the potential of multilateral printing inenergy applications	3D-printed parts may be used in electrical or electronic applications. Electrical properties like conductivity, resistivity, or dielectric constant can be evaluated using techniques like four-point probe or impedance spectroscopy	(Schwartz et al., 2022)

(Continued)

TABLE 3.1 (CONTINUED)

Important Characterization of AM-Produced Parts and Post Analysis

Characteristics	Objects	Process	Materials	Function	Remarks	Ref.
Biocompatibility	Bio-inspired bone scaffold	Digital Light Processing (DLP)	ZrO_2 powder, photosensitive resin	Scaffolds require good biocompatibility and moderate mechanical properties	3D-printed parts may be used in medical or biological applications where biocompatibility is important. Techniques like cell culture or in vitro testing can be used to assess the biocompatibility of the printed parts.	(Braconnier et al., 2020)
Environmental testing	Thin-walled apple part	Hand extrusion and casting, followed by 3D printing	Novel compostable/ biodegradable	Lower embodied impacts and being compostable or biodegradable	3D-printed parts may be exposed to different environmental conditions, such as temperature, humidity, or UV radiation. Environmental testing can be used to evaluate the durability and stability of the printed parts under different conditions	(Ratsimba et al., 2021)
Fatigue testing	Experimental specimen	Selective Laser Melting (SLM)	13H steel	Life cycle of the 3D-printed metallic specimen with mechanical properties, surface metrology fractography, and micro-CT investigated and reported	3D-printed parts may be subjected to repeated loading or cyclic stress, which can lead to fatigue failure. Fatigue testing can be used to evaluate the fatigue life and durability of the printed parts under cyclic loading	(Macek et al., 2022)

(Continued)

TABLE 3.1 (CONTINUED)

Important Characterization of AM-Produced Parts and Post Analysis

Characteristics	Objects	Process	Materials	Function	Remarks	Ref.
Rheological properties	Green samples for experimentation	Screw extruder and the printing nozzle, with sintering - 1350 °C for 2	3D gel-printing (3DGP), 316L stainless steel	Rheological behaviour of metal slurry	The flow behaviour and viscosity of the 3D-printing material can affect the quality and accuracy of the printed part. Rheological analysis techniques, such as melt flow index testing or rheometry, can be used to evaluate the flow properties of the printing material	(Du Plessis et al., 2016)
Adhesion strength	Testing specimen	FDM process	Polyvinyl alcohol (PVA), a water-soluble polymer, PLA mixed with metal (aluminium, copper) or carbon fibre	Adhesion of metal coatings on polymers	The adhesion strength between the printed layers can affect the mechanical properties and overall integrity of the printed part. Adhesion testing, such as peel or shear testing, can be used to evaluate the strength of the interlayer bonding.	(Ramaraju & Chandra, 2022)
Colour analysis	Step sample with induced layered defects	Powder-Bed Laser Additive Manufacturing	Ti6Al4V-titanium-6aluminum-4vanadium	Medical and aerospace industries	In applications where the colour or appearance of the printed part is important, colour analysis techniques can be used to evaluate colour accuracy and consistency	(Faludi et al., 2019)

(Continued)

TABLE 3.1 (CONTINUED)

Important Characterization of AM-Produced Parts and Post Analysis

Characteristics	Objects	Process	Materials	Function	Remarks	Ref.
Acoustic properties	PPM samples with different perforation angles were printed by using a 3D printer.	Stereolithography (SLA) by ProJet 7000	Porous polycarbonate material	Airgap investigation	3D-printed parts may be used in applications where acoustic properties, such as sound absorption or transmission, are important. Acoustic testing techniques, such as impedance tube or sound intensity measurement, can be used to evaluate these properties.	(Macek et al., 2022)
Optical properties	Dense metal of complex structural forms in an open-air environment using off-the-shelf metal rods as feedstock	Extrusion-based additive manufacturing and FED	6061 aluminium alloys	Direct deposition of fine-grained	3D-printed parts may be used in applications where optical properties, such as transparency or reflectivity, are important. Techniques like spectrophotometry or ellipsometry can be used to evaluate these properties.	(Gao et al., 2015)

associated with their development and application. The use of advanced materials has enabled the creation of complex and functional products in various industries, such as aerospace, automotive, and medical. However, to fully exploit the potential of additive manufacturing, it is crucial to continue developing new advanced materials with improved properties, processing techniques, and lower costs. With the continued progress in materials science and additive manufacturing technology, we can expect to see exciting new applications and innovations in the future.

REFERENCES

Adeniran, O., Cong, W., Bediako, E., & Aladesanmi, V. (2021). Additive manufacturing of carbon fiber reinforced plastic composites: The effect of fiber content on compressive properties. *Journal of Composites Science*, 5(12), 325. 10.3390/jcs5120325

Agarwal, S., Curtin, J., Duffy, B., & Jaiswal, S. (2016). Biodegradable magnesium alloys for orthopaedic applications: A review on corrosion, biocompatibility and surface modifications. *Materials Science and Engineering: C*, 68, 948–963. 10.1016/j.msec.2016.06.020

Al-Furjan, M. S. H., Shan, L., Shen, X., Zarei, M. S., Hajmohammad, M. H., & Kolahchi, R. (2022). A review on fabrication techniques and tensile properties of glass, carbon, and Kevlar fiber reinforced rolymer composites. *Journal of Materials Research and Technology*, 19, 2930–2959. 10.1016/j.jmrt.2022.06.008

Bahnini, I., Rivette, M., Rechia, A., Siadat, A., & Elmesbahi, A. (2018). Additive manufacturing technology: The status, applications, and prospects. *The International Journal of Advanced Manufacturing Technology*, 97, 147–161. 10.1007/s00170-018-1932-y

Bhardwaj, T., Shukla, M., Paul, C. P., & Bindra, K. S. (2019). Direct energy deposition-laser additive manufacturing of titanium-molybdenum alloy: Parametric studies, microstructure and mechanical properties. *Journal of Alloys and Compounds*, 787, 1238–1248. 10.1016/j.jallcom.2019.02.121

Blakey-Milner, B., Gradl, P., Snedden, G., Brooks, M., Pitot, J., Lopez, E., Bereto, F., Leary, M., & Du Plessis, A. (2021). Metal additive manufacturing in aerospace: A review. *Materials & Design*, 209, 110008. 10.1016/j.matdes.2021.110008

Braconnier, D. J., Jensen, R. E., & Peterson, A. M. (2020). Processing parameter correlations in material extrusion additive manufacturing. *Additive Manufacturing*, 31, 100924. 10.1016/j.addma.2019.100924

Du Plessis, A., le Roux, S. G., Booysen, G., & Els, J. (2016). Directionality of cavities and porosity formation in powder-bed laser additive manufacturing of metal components investigated using X-ray tomography. *3D Printing and Additive Manufacturing*, 3(1), 48–55. 10.1089/3dp.2015.0034

Engerroff, J. A. B., Baldissera, A. B., Magalhães, M. D., Lamarão, P. H., Wendhausen, P. A. P., Ahrens, C. H., & Mascheroni, J. M. (2019). Additive manufacturing of Sm-Fe-N magnets. *Journal of Rare Earths*, 37(10), 1078–1082. 10. 1078-1082. 10.1016/j.jre.2019.04.012

Ewais, E. M. M., Attia, M. A. A., Abousree-Hegazy, A., & Bordia, R. K. (2010). Investigation of the effect of ZrO2 and ZrO2/Al2O3 additions on the hot-pressing and properties of equimolecular mixtures of α-and β-Si3N4. *Ceramics International*, 36(4), 1327–1338. 10.1016/j.ceramint.2010.01.018

Faludi, J., Van Sice, C. M., Shi, Y., Bower, J., & Brooks, O. M. (2019). Novel materials can radically improve whole-system environmental impacts of additive manufacturing. *Journal of Cleaner Production*, 212, 1580–1590. 10.1016/j.jclepro.2018.12.017

Gao, W., Zhang, Y., Ramanujan, D., Ramani, K., Chen, Y., Williams, C. B., Wang, C. L. C., Shin, Y. C., Zhang, S., & Zavattieri, P. D. (2015). The status, challenges, and future of additive manufacturing in engineering. *Computer-Aided Design*, *69*, 65–89. 10.1016/j.cad.2015.04.001

Gibson, I., Rosen, D. W., Stucker, B., Gibson, I., Rosen, D. W., & Stucker, B. (2010). Sheet lamination processes. *Additive Manufacturing Technologies: Rapid Prototyping to Direct Digital Manufacturing*, 223–252. 10.1007/978-1-4419-1120-9_8

Kalman, H., Tripathi, N. M., Gabrieli, O. G., & Portnikov, D. (2017). Phase diagrams for pneumatic and hydraulic conveying. In *18th International Conferences on Transport and Sedimentation of Solid Particles, T and S 2017* (pp. 145–152). Wydawnictwo Uniwersytetu Przyrodniczego we Wroclawiu.

Gokuldoss, P. K., Kolla, S., & Eckert, J. (2017). Additive manufacturing processes: Selective laser melting, electron beam melting and binder jetting—Selection guidelines. *materials*, *10*(6), 672. 10.3390/ma10060672

Gülcan, O., Günaydın, K., & Tamer, A. (2021). The state of the art of material jetting—A critical review. *Polymers*, *13*(16), 2829. 10.3390/polym13162829

Huang, Y., Leu, M. C., Mazumder, J., & Donmez, A. (2015). Additive manufacturing: Current state, future potential, gaps and needs, and recommendations. *Journal of Manufacturing Science and Engineering*, *137*(1), 014001. 10.1115/1.4028725

Lakhdar, Y., Tuck, C., Binner, J., Terry, A., & Goodridge, R. (2021). Additive manufacturing of advanced ceramic materials. *Progress in Materials Science*, *116*, 100736. 10.1016/j.pmatsci.2020.100736

Laureto, J., Tomasi, J., King, J. A., & Pearce, J. M. (2017). Thermal properties of 3-D printed polylactic acid-metal composites. *Progress in Additive Manufacturing*, *2*, 57–71. 10.1007/s40964-017-0019-x

Li, S., Bai, H., Shepherd, R. F., & Zhao, H. (2019). Bio-inspired design and additive manufacturing of soft materials, machines, robots, and haptic interfaces. *Angewandte Chemie International Edition*, *58*(33), 11182–11204. 10.1002/anie.201813402

Lumay, G., Tripathi, N. M., & Francqui, F. (2019). How to gain a full understanding of powder flow properties, and the benefits of doing so. *ONdrugDelivery*, *2019*, 42–47.

Macek, W., Martins, R. F., Branco, R., Marciniak, Z., Szala, M., & Wroński, S. (2022). Fatigue fracture morphology of AISI H13 steel obtained by additive manufacturing. *International Journal of Fracture*, *235*(1), 79–98. 10.1007/s10704-022-00615-5

M Tripathi, N., & S Mallick, S. (2017). Pneumatic conveying of Fly Ash: Bend Models investigation. *Advanced Materials Proceedings*, *2*(8), 526–531.

Neveu, A., Lumay, G., Pillitteri, S., Monsuur, F., Pauly, T., Ribeyre, Q., Francqui, F., Vandewalle, N., & Tripathi, N. M. (2020b). Physical characterization of blends containing mesoporous particles with a focus on electrostatic properties. In *2020 AIChE Spring Meeting and 16th Global Congress on Process Safety*. AIChE.

Neveu, A., Tripathi, N. M., Rigo, O., Francqui, F., & Lumay, G. (2020a). Experimental investigation of spreadability of metal powders in recoating process. In *Proceedings - Euro PM2020 Congress and Exhibition (Proceedings - Euro PM2020 Congress and Exhibition)*. European Powder Metallurgy Association (EPMA).

Ramaraju, R. V., & Chandra, S. (2022). Additive manufacturing of metal components by thermal spray deposition on 3D-printed polymer parts. *Journal of Thermal Spray Technology*, *31*(8), 2409–2421. 10.1007/s11666-022-01450-9

Ren, X., Shao, H., Lin, T., & Zheng, H. (2016). 3D gel-printing—An additive manufacturing method for producing complex shape parts. *Materials & Design*, *101*, 80–87. 10.1016/j.matdes.2016.03.152

Saleh Alghamdi, S., John, S., Roy Choudhury, N., & Dutta, N. K. (2021). Additive manufacturing of polymer materials: Progress, promise and challenges. *Polymers*, *13*(5), 753. 10.3390/polym13050753

Schwartz, J. J. (2022). Additive manufacturing: Frameworks for chemical understanding and advancement in vat photopolymerization. *MRS Bulletin*, *47*(6), 628–641. 10.1557/s43577-022-00343-0

Shah, J., Snider, B., Clarke, T., Kozutsky, S., Lacki, M., & Hosseini, A. (2019). Large-scale 3D printers for additive manufacturing: Design considerations and challenges. *The International Journal of Advanced Manufacturing Technology*, *104*(9–12), 3679–3693. 10.1007/s00170-019-04074-6

Sharma, A., Babbar, A., Tian, Y., Pathri, B. P., Gupta, M., & Singh, R. (2022). Machining of ceramic materials: A state of the art review. *International Journal on Interactive Design and Manufacturing (IJIDeM)*, 1–21. https://doi.org/10.1007/s12008-022-01016-7

Sharma, A., & Jain, V. (2020). Experimental investigation of cutting temperature during drilling of float glass specimen. In *IOP Conference Series: Materials Science and Engineering* (Vol. 715, No. 1, p. 012050). IOP Publishing.

Sharma, A., Jain, V., & Gupta, D. (2018). Characterization of chipping and tool wear during drilling of float glass using rotary ultrasonic machining. *Measurement*, *128*, 254–263.

Sharma, A., Jain, V., & Gupta, D. (2019a). Tool wear analysis while creating blind holes on float glass using conventional drilling: A multi-shaped tools study. In *Advances in Manufacturing Processes: Select Proceedings of ICEMMM 2018* (pp. 175–183). Springer Singapore.

Sharma, A., Jain, V., & Gupta, D. (2019b). Comparative analysis of chipping mechanics of float glass during rotary ultrasonic drilling and conventional drilling: For multi-shaped tools. *Machining Science and Technology*, *23*(4), 547–568.

Sharma, A., Jain, V., & Gupta, D. (2021a). Effect of pre and post tempering on hole quality of float glass specimen: For rotary ultrasonic and conventional drilling. *Silicon*, *13*, 2029–2039.

Sharma, A., Jain, V., Gupta, D., & Babbar, A. (2020). A review study on miniaturization. In *Advanced Manufacturing and Processing Technology* (First edition, pp. 111–131). CRC Press.

Snow, Z., Nassar, A. R., & Reutzel, E. W. (2020). Invited Review Article: Review of the formation and impact of flaws in powder bed fusion additive manufacturing. *Additive Manufacturing*, *36*, 101457. 10.1016/j.addma.2020.101457

Tripathi, N. M., Francqui, F., & Lumay, G. (2020). Influence of relative air humidity on the flow property of fine powders. In *Third International Conference on Powder, Granule and Bulk Solids: Innovations and Applications PGBSIA 2020 February 26–28, 2020* (p. 63).

Tripathi, N. M., Francqui, F., Pirenne, T., & Lumay, G. (2019). Measuring food powders electrical properties as a result of anti-static content. In *2019 AIChE Annual Meeting*. American Institute of Chemical Engineers.

Tripathi, N. M., Levy, A., & Kalman, H. (2016). Initial acceleration pressure drop in dilute phase pneumatic conveying system. In *Powder, Granule and Bulk Solids: Innovations and Applications Conference*.

Tripathi, N. M., & Mallick, S. S. (2014). *An Investigation into Pressure Drop Across Bends for Fluidised Densephase Pneumatic Conveying Systems* (Doctoral dissertation).

Vyas, D., & Udyawar, D. (2019). A review on current state of art of bioprinting. *3D Printing and Additive Manufacturing Technologies*, 195–201. 10.1007/978-981-13-0305-0_17

Zhang, Y., Wu, L., Guo, X., Kane, S., Deng, Y., Jung, Y. G., … Zhang, J. (2018). Additive manufacturing of metallic materials: A review. *Journal of Materials Engineering and Performance*, *27*, 1–13. 10.1007/s11665-017-2747-y

4 Investigations on the Mechanical Characterization of Additively Manufactured Polymer Composite Materials Fabricated Via MEX

Akant Kumar Singh, Mufaddal Huzefa Shakir, Siddhartha, Sanjay Yadav, Manvendra Yadav, Naveen Mani Tripathi, and Ismail Fidan

4.1 INTRODUCTION

Additive Manufacturing (AM), also known as 3D Printing (3DP), has surfaced as a groundbreaking production technology, bringing about a revolution in numerous industries, including highly demanding sectors such as aerospace and biomedical industries (Fidan et al., 2023). 3DP is a technique for creating objects by layering together specific materials as needed under the direction of a digital 3D model. The industry, as well as the academic and research groups, have shown an increased interest in 3DP. Recently, faster, less-expensive AM processes that can achieve excellent print quality have been created. Additionally, a greater variety of polymer materials with different qualities are also being manufactured for 3DP. These developments continuously alter the way things are created, used by consumers, and how they are developed and made (Al Rashid et al., 2020; Siviour et al., 2016). Because 3DP makes producing prototypes so much easier, innovators and inventors can now easily test their ideas. In fact, the design and manufacturing processes have been sped up from weeks to a few hours, thereby enabling on-the-fly innovation (Shanmugam et al., 2021). AM could reduce production costs and increase overall industrial efficiency (Ngo et al., 2018). Additionally, AM provides quick turnaround times and

DOI: 10.1201/9781003428862-4

small batch sizes for applications where complicated designs are needed. Several applications, including those in construction, fashion, dentistry, medicine, electronics, automotive, robots, military, oceanographic, aerospace, and others are currently being seriously examined for AM-produced materials (Baumann et al., 2017). Few studies have also discussed the investigations on the characterization of additively manufactured materials (Sharma et al., 2022; Sharma & Jain, 2020; Sharma et al., 2018; Sharma et al., 2019a; Sharma et al., 2019b; Sharma et al., 2019c; Sharma et al., 2021a; Sharma et al., 2021b; Sharma et al., 2020; Sharma et al., 2022; Kalman et al., 2017; Lumay et al., 2019; Tripathi & Mallick, 2017; Neveu et al., 2020a; Neveu et al., 2020b; Ratsimba et al., 2021; Tripathi et al., 2020; Tripathi et al., 2019; Tripathi et al., 2016; Tripathi & Mallick, 2014).

Materials made of polymers are frequently employed in the 3DP sector owing to their low cost, less weight, and flexibility in processing. The properties of polymer matrix composites can be greatly enhanced by using fibers as reinforcements (Wang et al., 2017). MEX is an often-utilized 3DP technique to create polymer composites reinforced with fibers (Gupta et al., 2023). To create filaments for the MEX process, polymer granules and fibers must first be combined in a blender before being fed to an extruder. A second extrusion procedure could be employed to guarantee that the fibers are distributed uniformly (Guo et al., 2013). To enrich the mechanical qualities of polymer composites in the field of 3DP, standard short fibers like glass and carbon are frequently utilized as reinforcements (Zhong et al., 2001; Ning et al., 2017). The characteristics of the finished composite products are significantly influenced by the void fraction and fiber orientation of composites (Love et al., 2014). Various polymer matrix composites have been manufactured using 3DP techniques (Fidan et al., 2019). A study looked into the effects of fiber arrangement and permeability on the characteristics of Carbon fiber/ABS composite specimens printed via MEX process. Comparatively, compression molding was used to create composite products. In contrast to compression molded samples, which had essentially no pores, composite specimens generated by 3DP had a formation of voids (20%) because of holes between deposition lines and poor interfacial adhesion between the fiber and polymer matrix. The improvement in tensile strength for the printed composite samples was comparable to that for compression molded specimens. This was so that the negative effects of porosity were offset by printing, which aligns more fibers in the direction of load bearing (Tekinalp et al., 2014). The composite with thermoplastic polymer and reinforcement of continuous fiber was fabricated using a Mark One printer, and its mechanical characteristics were investigated. The printed sample featured a sandwich architecture, with the top and the bottom layer comprising nylon polymer and carbon fiber-reinforced thermoplastic polymer in the center. To extrude CFRP and nylon, two print heads were used, respectively (Van Der Klift et al., 2016). According to a study, the ultimate tensile strength and elastic modulus of PLA composites reinforced with continuous carbon fiber were 185.2 MPa and 19.5 GPa which were 435% and 599% of the tensile strength and modulus of pure PLA samples. Comparing this mechanical improvement to that of fiber-reinforced PLA composites, it was substantially larger (Matsuzaki et al., 2016). In this study, the mechanical characteristics, like tensile, bending, and compressive strength, of composites created using Onyx as the matrix and fiberglass as reinforcement have been evaluated. Composite specimens were created using the MEX technique and a Mark Two 3D printer. The results of the fabricated composites are addressed in depth in this chapter.

4.2 MATERIALS AND METHODS

4.2.1 MATERIALS

Onyx matrix was purchased from Markforged. Onyx fabricates components with a rigidity that is on par with or better than any pure thermoplastic material now offered for use in professional 3D printers. It is much more rigid and easy to print and assemble. Onyx can be employed alone or with implanted continuous Kevlar, carbon fiber, and fiberglass. Fiberglass is light, robust, and brittle material. Fiberglass can be shaped into a variety of forms. Table 4.1 depicts the properties of Onyx and fiberglass.

4.2.2 COMPOSITE PREPARATION

The composite specimens were prepared using a Mark Two 3D printer with different compositions. The specimens were developed as per the standard shape and size required for different mechanical testing. In this chapter, a single nominal layer thickness of 0.2 mm was assessed. Only 20 layers with an ostensible thickness of 0.2 mm make up the tensile and bending samples, whereas 200 layers with an ostensible thickness of 0.2 mm make up the compressive test's sample. Each specimen's actual thickness and width were measured with a caliper, and minimum measurements were used for computations in compliance with testing standards. A total of three samples were evaluated for each test. Neat Onyx, fiberglass composite, and Onyx + fiberglass composites were fabricated in this experiment. Sampling details of the specimens are shown in Table 4.2.

TABLE 4.1
Properties of Fiberglass and Onyx (Bárnik et al., 2019; Çuvalci et al. 2014)

Parameters	Onyx	Fiberglass
Density (g/cm3)	1.2	2.56
Tensile strength (MPa)	36	3,445
Elastic modulus (GPa)	1.4	76
Bending strength (MPa)	81	200
Bending modulus (GPa)	0.0029	22
Compressive strength (MPa)	–	1,080
Tensile strain at break (%)	58	2.75
Flexural strain at break (%)	–	1.1
Compressive strain at break (%)	–	–

TABLE 4.2
Sampling Preparation Details for Various Testing

Composite	Print Time (hr)	Onyx Volume (cc)	Fiber Volume (cc)
Tensile (Onyx)	1	6.81	–
Tensile (Fiberglass)	2	3.34	3.5
Tensile (Onyx + Fiberglass)	1.5	3.56	3.1
Compressive (Onyx)	2	12.8	–
Compressive (Fiberglass)	6	6.2	8.88
Compressive (Onyx + Fiberglass)	5.1	5.7	7.2
Bending (Onyx)	1	5.98	–
Bending (Fiberglass)	2.8	2.2	3.98
Bending (Onyx + Fiberglass)	2.4	2.45	3.83

4.2.3 MECHANICAL TESTS

The mechanical characteristics like bending, tensile, and compressive strength of the fabricated composites are assessed in this study. Tensile testing is a destructive test that reveals details about the mechanical characteristics of a material, including yield strength, tensile strength, and ductility. With this kind of testing, the amount of elongation required to break a specimen as well as the required force are both determined. Compression test is used to evaluate a material or product's strength and stiffness under crushing loads. Bending test determines a material's capacity to endure deformation under load and reveals the material's maximum stress at the point of rupture (Anand et al., 2019; Kumar et al., 2022).

Tensile and bending tests are conducted in accordance with ASTM D638 standards. For tensile, compressive, and bending tests, three types of materials are printed: one with neat Onyx, another with Onyx + fiberglass, and a third with only fiberglass. While the Compression Testing Machine (CTM) is used for compression tests, the Universal Testing Machine (UTM) is utilized for tensile and bending tests. The experiment was conducted at room temperature. Figures 4.1 and 4.2 show the composite samples placed in UTM and CTM.

4.3 RESULTS AND DISCUSSION

4.3.1 TENSILE TEST

Tensile testing on both the Onyx and Onyx + fiberglass specimens was completed successfully. The results are presented in detail below in the following sections.

4.3.1.1 Neat Onyx

The composite specimen was subjected to tensile loading in UTM as seen in Figure 4.3. The specimen was placed in UTM and loaded until the fracture point was observed. A load versus deformation curve was plotted using the data, as seen

FIGURE 4.1   Specimen placed in UTM for the tensile test.

FIGURE 4.2   Specimen placed in CTM for the compressive test.

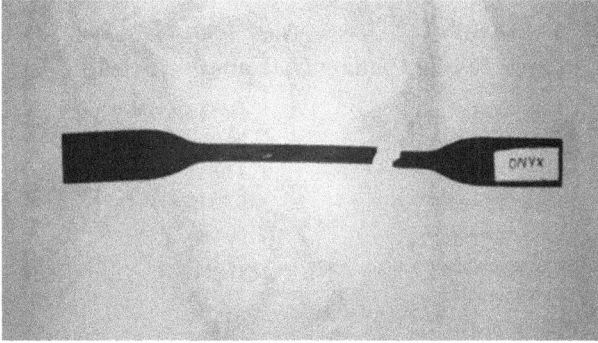

FIGURE 4.3   Neat Onyx specimen after tensile loading in UTM.

in Figure 4.4. It was evident from the displayed curve that the material experiences tensile failure at a maximum stress of 3.0220 KN with a maximum deformation of 39.8069 mm. According to the measurements, the material's elastic modulus was 0.78 GP. Onyx alone has a very encouraging tensile strength of 126 KN. Table 4.3. depicts Onyx tensile loading data and the results.

4.3.1.2   Onyx + Fiberglass

Figure 4.5 displays the composite specimen after tensile loading in UTM. Onyx + fiberglass tensile loading data and the results are summarized in Table 4.4. According to the curve depicted in Figure 4.6, the material's tensile failure occurs at a maximum loading of 8.088 KN and a maximum deformation of 6.6512 mm. The specimen had an elastic modulus of 4.54 GPa. The composite created with Onyx + fiberglass has a 335 KN tensile strength, which is quite encouraging.

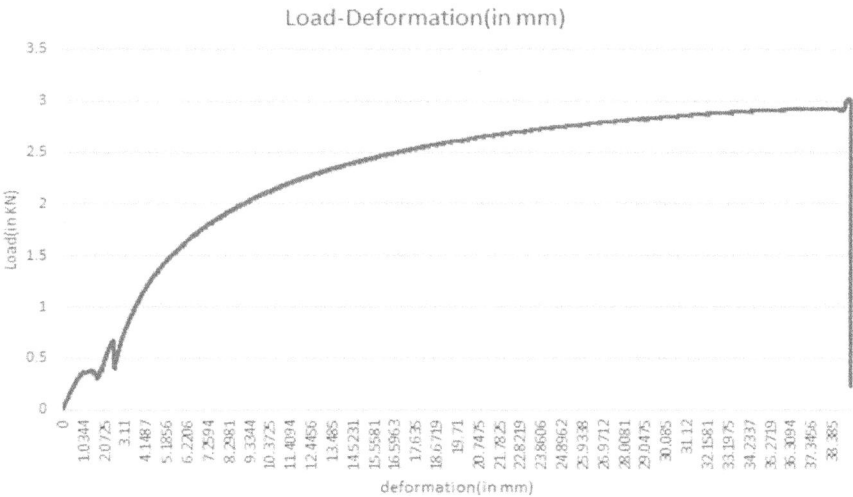

FIGURE 4.4   Load versus deformation curve of neat Onyx specimen.

TABLE 4.3
Onyx Tensile Loading Data and the Results

Parameters	Onyx specimen
Area	24 mm
Size	150*16*4 (mm)
Layer thickness	0.2 μm
Gauge length	60 mm
Maximum deformation	39.8069 mm
Maximum load	3.0220 KN
Elastic modulus (GPa)	0.78
Tensile strength (KN)	126.0

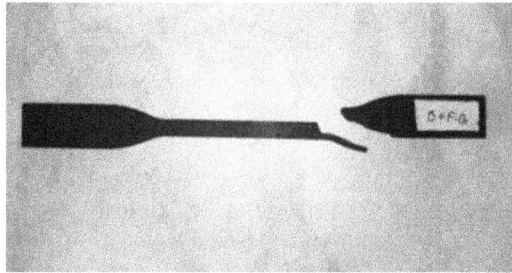

FIGURE 4.5 Onyx + fiberglass specimen after tensile loading in UTM.

TABLE 4.4
Onyx + Fiberglass Tensile Loading Data and the Results

Parameters	Onyx + Fiberglass specimen
Area	24 mm
Size	150*16*4 (mm)
Layer thickness	0.2 μm
Gauge length	60 mm
Maximum deformation	6.6152 mm
Maximum load	8.088 KN
Elastic modulus (GPa)	4.54
Tensile strength (KN)	335

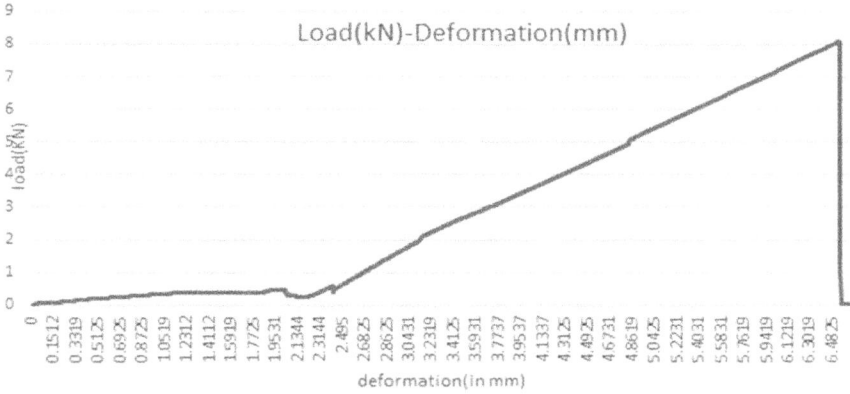

FIGURE 4.6 Load versus deformation curve of Onyx + fiberglass specimen.

4.3.2 COMPRESSION TEST

The results obtained following the successful completion of compressive testing on all three specimens (Onyx, fiberglass, and Onyx + fiberglass) are covered in more detail below.

4.3.2.1 Neat Onyx

A CTM was used to conduct compression testing on the sample. Figure 4.7 shows the neat Onyx composite sample after the compression test. After the information was gathered, it was clear that the material deforms when subjected to a maximum loading of 349 KN.

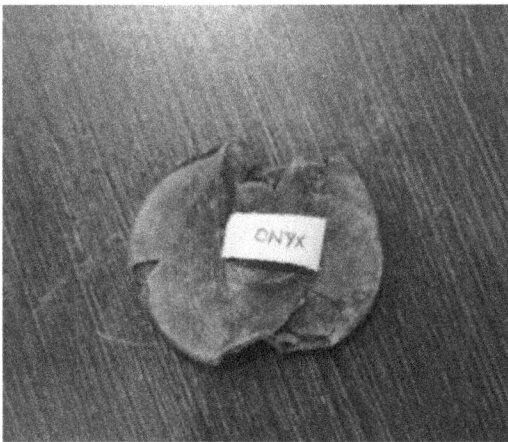

FIGURE 4.7 Neat Onyx specimen after compressive loading in CTM.

FIGURE 4.8 Fiberglass specimen after compressive loading in CTM.

4.3.2.2 Fiberglass

The sample was compressed in a CTM. To the point of fracture, the specimen was loaded in the CTM. Analysis of the data revealed that the material compresses at a maximum force of 81 KN, leading to the deformation shown in Figure 4.8.

4.3.2.3 Onyx + Fiberglass

The sample was compressed in a compression-testing equipment. The specimen was loaded in the CTM until it broke. After analyzing the data, it was found that the material compresses at a maximum load of 61 KN, leading to the deformation shown in Figure 4.9. Table 4.5 shows the summary of compression test results.

4.3.3 BENDING TEST

The bending test on all three fabricated composites was performed successfully. The results of bending tests are explained in detail below.

FIGURE 4.9 Onyx + fiberglass specimen after compressive loading in CTM.

TABLE 4.5
Summary of All the Three Specimens Subjected to Compressive Load

Specimen	Maximum load
Onyx	349 KN
Fiberglass	81 KN
Onyx + Fiberglass	61 KN

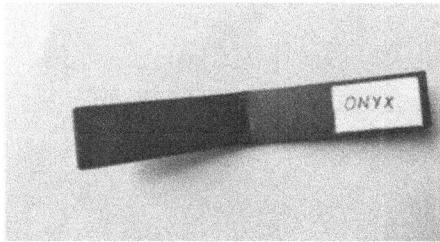

FIGURE 4.10 Neat Onyx specimen after the bending test.

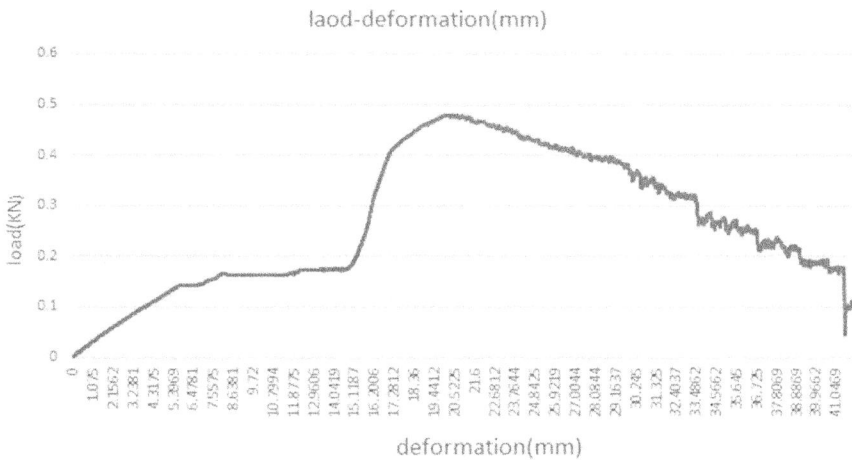

FIGURE 4.11 Load versus deformation curve of neat Onyx specimen after bending test.

4.3.3.1 Neat Onyx

The bending was conducted on a UTM machine. Figure 4.10 shows the neat Onyx sample after being subjected to the bending test. According to the illustrated curve in Figure 4.11, the specimen bends at a maximum loading of 0.4799 KN and a maximum deformation of 20.6525 mm.

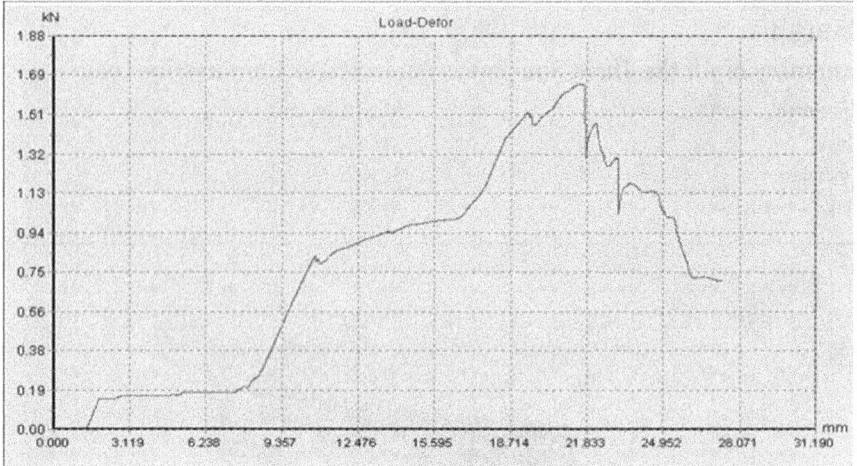

FIGURE 4.12   Load versus deformation curve of fiberglass specimen after bending test.

4.3.3.2   Fiberglass

Figure 4.12 shows how the specimen bends at a maximum load of 1.652 KN and a maximum deformation of 21.833 mm during the compression test.

4.3.3.3   Onyx + Fiberglass

As shown in Figure 4.13, the specimen bends at a maximum loading of 1.156 KN and a maximum deformation of 20.5 mm.

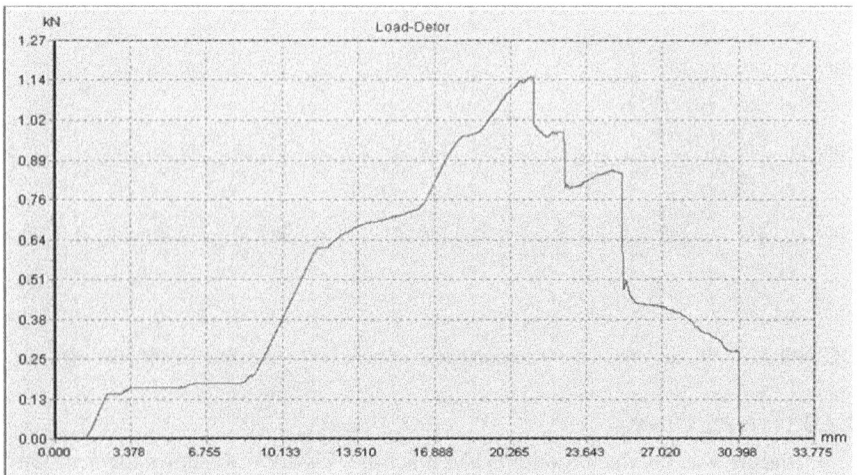

FIGURE 4.13   Load versus deformation curve of Onyx + Fiberglass specimen after bending test.

4.4 CONCLUSION

This study generates a number of knowledge blocks about the practical utilization of continuous fiber-reinforced composite components fabricated with Onyx, fiberglass, and glass fiber. It is believed that the information generated will be used by a number of the practitioners in engineering and technology fields manufacturing products with the continuous fiber-reinforced MEX process. The results of the fabricated composites are summarized below:

- In the tensile test, it is observed that neat Onyx has shown more ductile behavior as compared to Onyx + fiberglass composite. On the contrary, Onyx + fiberglass composite has more load-bearing capacity. Onyx + fiberglass composite can bear 30% more load than neat Onyx.
- From the results of the compressive test, it was found that neat Onyx is 76.40% stronger than glass fiber composite and 82.52% stronger than the composite made of Onyx + fiberglass. It may be concluded that additively manufactured Onyx + fiberglass composite has suffered a loss in compressive strength.
- In the bending test, it was observed that neat Onyx was capable of tolerating a maximum load of 0.48 KN, fiberglass composite was capable of tolerating a maximum load of 1.65 KN, and Onyx + fiberglass composite was able to endure a maximum load of 1.15 KN. Therefore, it may be said that fiberglass composite has the highest bending strength as compared to neat Onyx and Onyx + fiberglass composite.

REFERENCES

Al Rashid, A., Khan, S.A., Al-Ghamdi, S.G., & Koç, M. (2020). Additive manufacturing: Technology, applications, markets, and opportunities for the built environment. *Automation in Construction*, *118*, 103268.

Anand, K.S., & Shivraj, N.Y. (2019). Tensile testing and evaluation of 3D-printed PLA specimens as per ASTM D638 type IV standard. In *Innovative Design, Analysis and Development Practices in Aerospace and Automotive Engineering (I-DAD 2018) Volume 2* (pp. 79–95). Springer Singapore.

Bárnik, F., Vaško, M., Handrik, M., Dorčiak, F., & Majko, J. (2019). Comparing mechanical properties of composites structures on Onyx base with different density and shape of fill. *Transportation Research Procedia*, *40*, 616–622.

Baumann, F.W., & Roller, D. (2017). Additive manufacturing, cloud-based 3D printing and associated services—Overview. *Journal of Manufacturing and Materials Processing*, *1*(2), 15.

Çuvalci, H., Erbay, K., & İpek, H. (2014). Investigation of the effect of glass fiber content on the mechanical properties of cast polyamide. *Arabian Journal for Science and Engineering*, *39*, 9049–9056.

Fidan, I., Huseynov, O., Alshaikh Ali, M., Alkunte, S., Rajeshirke, M., Gupta, A., Hasanov, S., Tantawi, K., Yasa, E., Yilmaz, O., Loy, J., Popov, V., & Sharma, A. (2023). Recent inventions in additive manufacturing: Holistic review. *Inventions*, *8*(4), 103. https://doi.org/10.3390/INVENTIONS8040103

Fidan, I., Imeri, A., Gupta, A., Hasanov, S., Nasirov, A., Elliott, A., Alifui-Segbaya, F., & Nanami, N. (2019). The trends and challenges of fiber reinforced additive manufacturing. *International Journal of Advanced Manufacturing Technology, 102*(5–8), 1801– 1818. https://doi.org/10.1007/S00170-018-03269-7

Guo, N., & Leu, M.C. (2013). Additive manufacturing: Technology, applications and research needs. *Frontiers of Mechanical Engineering, 8*, 215–243.

Gupta, A., Hasanov, S., Alifui-Segbaya, F., & Fidan, I. (2023). Composites (Fiber-reinforced plastic matrix composites). In *Springer Handbook of Additive Manufacturing* (pp. 627– 637). Springer International Publishing. https://doi.org/10.1007/978-3-031-20752-5_37

Kalman, H., Tripathi, N. M., Gabrieli, O. G., & Portnikov, D. (2017). Phase diagrams for pneumatic and hydraulic conveying. In *18th International Conferences on Transport and Sedimentation of Solid Particles, T and S 2017* (pp. 145–152). Wydawnictwo Uniwersytetu Przyrodniczego we Wroclawiu.

Kumar, S., Singh, R., Singh, T.P., & Batish, A. (2022). On investigations of thermal conductivity, circumferential compressive strength, and surface characterization of 3D-printed hybrid blended magnetostrictive PLA composite. *Journal of Thermoplastic Composite Materials, 35*(5), 631–650.

Love, L.J., Kunc, V., Rios, O., Duty, C.E., Elliott, A.M., Post, B.K., Smith, R.J., & Blue, C.A. (2014). The importance of carbon fiber to polymer additive manufacturing. *Journal of Materials Research, 29*(17), 1893–1898.

Lumay, G., Tripathi, N. M., & Francqui, F. (2019). How to gain a full understanding of powder flow properties, and the benefits of doing so. *ONdrugDelivery, 2019*, 42–47.

Matsuzaki, R., Ueda, M., Namiki, M., Jeong, T.K., Asahara, H., Horiguchi, K., Nakamura, T., Todoroki, A., & Hirano, Y. (2016). Three-dimensional printing of continuous-fiber composites by in-nozzle impregnation. *Scientific Reports, 6*(1), 23058.

M Tripathi, N., & S Mallick, S. (2017). Pneumatic conveying of Fly Ash: Bend Models investigation. *Advanced Materials Proceedings, 2*(8), 526–531.

Neveu, A., Lumay, G., Pillitteri, S., Monsuur, F., Pauly, T., Ribeyre, Q., Francqui, F., Vandewalle, N., & Tripathi, N. M. (2020b). Physical characterization of blends containing mesoporous particles with a focus on electrostatic properties. In *2020 AIChE Spring Meeting and 16th Global Congress on Process Safety*. AIChE.

Neveu, A., Tripathi, N. M., Rigo, O., Francqui, F., & Lumay, G. (2020a). Experimental investigation of spreadability of metal powders in recoating process. In *Proceedings - Euro PM2020 Congress and Exhibition (Proceedings - Euro PM2020 Congress and Exhibition)*. European Powder Metallurgy Association (EPMA).

Ngo, T.D., Kashani, A., Imbalzano, G., Nguyen, K.T., & Hui, D. (2018). Additive manufacturing (3D printing): A review of materials, methods, applications and challenges. *Composites Part B: Engineering, 143*, 172–196.

Ning, F., Cong, W., Hu, Y., & Wang, H. (2017). Additive manufacturing of carbon fiber-reinforced plastic composites using fused deposition modeling: Effects of process parameters on tensile properties. *Journal of Composite Materials, 51*(4), 451–462.

Ratsimba, A., Zerrouki, A., Tessier-Doyen, N., Nait-Ali, B., André, D., Duport, P., Neveu, A., Tripathi, N., Francqui, F., & Delaizir, G. (2021). Densification behaviour and three-dimensional printing of Y2O3 ceramic powder by selective laser sintering. *Ceramics International, 47*(6), 7465–7474. https://doi.org/10.1016/j.ceramint.2020.11.087

Shanmugam, V., Rajendran, D.J.J., Babu, K., Rajendran, S., Veerasimman, A., Marimuthu, U., Singh, S., Das, O., Neisiany, R.E., Hedenqvist, M.S., & Berto, F. (2021). The mechanical testing and performance analysis of polymer-fibre composites prepared through the additive manufacturing. *Polymer Testing, 93*, 106925.

Sharma, A., Babbar, A., Tian, Y., Pathri, B.P., Gupta, M., & Singh, R. (2022). Machining of ceramic materials: A state of the art review. *International Journal on Interactive Design and Manufacturing (IJIDeM)*, 1–21. https://doi.org/10.1007/s12008-022-01016-7

Sharma, A., & Jain, V. (2020). Experimental investigation of cutting temperature during drilling of float glass specimen. In *IOP Conference Series: Materials Science and Engineering* (Vol. 715, No. 1, p. 012050). IOP Publishing.

Sharma, A., Jain, V., & Gupta, D. (2018). Characterization of chipping and tool wear during drilling of float glass using rotary ultrasonic machining. *Measurement*, *128*. 254–263.

Sharma, A., Jain, V., & Gupta, D. (2019a). Tool wear analysis while creating blind holes on float glass using conventional drilling: A multi-shaped tools study. In *Advances in Manufacturing Processes: Select Proceedings of ICEMMM 2018* (pp. 175–183). Springer Singapore.

Sharma, A., Jain, V., & Gupta, D. (2019b). Comparative analysis of chipping mechanics of float glass during rotary ultrasonic drilling and conventional drilling: For multi-shaped tools. *Machining Science and Technology*, *23*(4), 547–568.

Sharma, A., Jain, V., & Gupta, D. (2019c). Multi-shaped tool wear study during rotary ultrasonic drilling and conventional drilling for amorphous solid. *Proceedings of the Institution of Mechanical Engineers, Part E: Journal of Process Mechanical Engineering*, *233*(3), 551–560.

Sharma, A., Jain, V., & Gupta, D. (2021a). Effect of pre and post tempering on hole quality of float glass specimen: For rotary ultrasonic and conventional drilling. *Silicon*, *13*, 2029–2039.

Sharma, A., Jain, V., & Gupta, D. (2021b). Mathematical approach on chipping volume estimation generated during rotary ultrasonic drilling for float glass. *Proceedings of the National Academy of Sciences, India Section A: Physical Sciences*, *92*, 285–291.

Sharma, A., Jain, V., Gupta, D., & Babbar, A. (2020). A review study on miniaturization. *Advanced Manufacturing and Processing Technology* (First edition, pp. 111–131). CRC Press.

Siviour, C.R., & Jordan, J.L. (2016). High strain rate mechanics of polymers: A review. *Journal of Dynamic Behavior of Materials*, *2*, 15–32.

Tekinalp, H.L., Kunc, V., Velez-Garcia, G.M., Duty, C.E., Love, L.J., Naskar, A.K., Blue, C.A., & Ozcan, S. (2014). Highly oriented carbon fiber–polymer composites via additive manufacturing. *Composites Science and Technology*, *105*, 144–150.

Tripathi, N. M., Francqui, F., & Lumay, G. (2020). Influence of relative air humidity on the flow property of fine powders. In *Third International Conference on Powder, Granule and Bulk Solids: Innovations and Applications PGBSIA 2020 February 26–28, 2020* (p. 63).

Tripathi, N. M., Francqui, F., Pirenne, T., & Lumay, G. (2019). Measuring food powders electrical properties as a result of anti-static content. In *2019 AIChE Annual Meeting*. American Institute of Chemical Engineers.

Tripathi, N. M., Levy, A., & Kalman, H. (2016). Initial acceleration pressure drop in dilute phase pneumatic conveying system. In *Powder, Granule and Bulk Solids: Innovations and Applications Conference*.

Tripathi, N. M., & Mallick, S. S. (2014). *An Investigation into Pressure Drop Across Bends for Fluidised Densephase Pneumatic Conveying Systems* (Doctoral dissertation).

Van Der Klift, F., Koga, Y., Todoroki, A., Ueda, M., Hirano, Y., & Matsuzaki, R. (2016). 3D printing of continuous carbon fibre reinforced thermo-plastic (CFRTP) tensile test specimens. *Open Journal of Composite Materials*, *6*(01), 18.

Wang, X., Jiang, M., Zhou, Z., Gou, J., & Hui, D. (2017). 3D printing of polymer matrix composites: A review and prospective. *Composites Part B: Engineering*, *110*, 442–458.

Zhong, W., Li, F., Zhang, Z., Song, L., & Li, Z. (2001). Short fiber reinforced composites for fused deposition modeling. *Materials Science and Engineering: A*, *301*(2), 125–130.

5 Investigations on Die-Swell Behaviour in Screw Extrusion-Based Processing of Thermoplastic ABS

Yash Gopal Mittal, Pushkar Kamble, Gopal Gote, Yogesh Patil, Avinash Kumar Mehta, and K. P. Karunakaran

5.1 INTRODUCTION

Polymeric material such as thermoplastic deforms under thermal loading conditions because of which thermoplastic polymers have a variety of uses in the engineering field (Olabisi & Adewale, 2016). Screw extrusion is a highly scalable method of thermoplastic processing, requiring pelletized feedstock (Reddy et al., 2007). In addition to direct industrial uses like *additive manufacturing* (AM) and *injection moulding* (IM), it is employed in other industries like food 3D printing (Voon et al., 2019).

The screw-extruder setup consists of various components, out of which the extruder screw is the most important. Its geometry can be dictated by a single parameter, i.e. the pitch (p), which also dominates the process characteristics. The pitch is a measure of the axial material flow per rotation of the screw and is defined with respect to the helix angle ($\varnothing$) and the diameter of the screw (D), as described in Equation 5.1. The extruder pitch, and the associated helix angle, should be meticulously designed, considering various geometrical and process parameters, and the subsequent application (Mittal et al., 2022).

$$\tan\varnothing = \frac{p}{\pi \times D} \qquad (5.1)$$

The extruder screw takes on a cylindrical shape and features spiral grooves on its outer surface, which guide the flow of molten polymer. This screw is situated within a heated barrel, equipped with a nozzle at its end. A heating unit, typically utilizing resistive elements, encircles the barrel to melt the thermoplastic material. The nozzle

DOI: 10.1201/9781003428862-5

serves both to converge the rotating molten material and to impart the final shape during extrusion. The process requires a motor-driven system characterized by low speed (between 10 RPM and 80 RPM) and high torque (10–30 Nm) (Das, 2008). The primary function of this power drive is to provide a steady rotational speed, as any variations can lead to changes in the dimensions of the extruded product.

The core operations in screw extrusion encompass solid conveying, melting, and melt pumping. The initial feeding zone of the screw receives solid polymer pellets, which are conveyed along its length. The heat from external heaters gradually melts the polymer, forming a thin layer that moves forward. In the transition zone, the plastic granules undergo melting and are pushed forward, ultimately transforming into a molten state. The final metering zone induces a circulating flow by skimming the polymer melt from the inner surface of the barrel and directing it downward along the flight, as depicted in Figure 5.1. The entire process is a combination of drag and pressure-driven flow, and it is customary to consider the effects of viscous drag and melt pressure in the flow analysis (Mount (2017)).

Because of the above-mentioned processing conditions of straining and temperature, the melt pressure generated inside the extruder is tremendous and can go up to 10,000 psi (Rosato, 2004). Therefore, screw extrusion can be considered as a *high-pressure* process. This tremendous melt pressure is experienced by the polymer melt, which is instantly released just after the extrusion at the nozzle. As a manifestation of this stress release, the extrusion undergoes free expansion, where the radial dimensions, i.e., the size of the extrusion, in terms of its diameter, increases. The ideal extrusion size should be equal to the size of the orifice at the nozzle end, but in practice, it is always greater than that. This phenomenon of free expansion of the extrusion, as a manifestation of stress release in the polymer melt, leading to an increased size of the extrudate, is called the die-swell, often denoted by (B). Die-swell is directly calculated by determining the ratio of the extrudate (d) to the nozzle size (d_N). Mathematically, die-swell is represented as (Equation 5.2):

$$B = \frac{d}{d_N} \qquad (5.2)$$

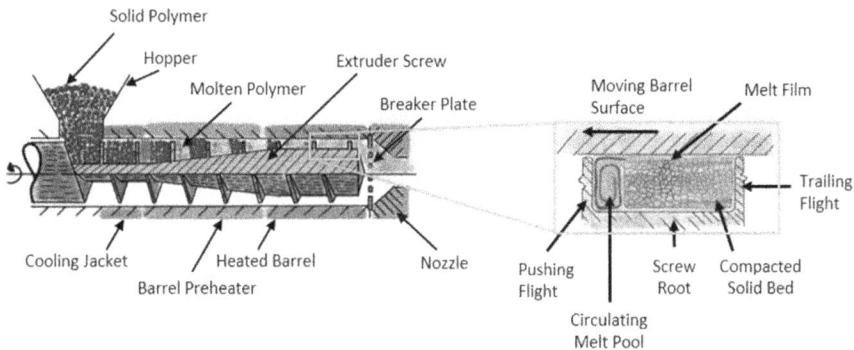

FIGURE 5.1 Section view of a single screw extruder.

It should be noted that, although the extruded material rapidly cools down after the free extrusion in the open environment, the associated volume contraction (because of the densification of the material) does not compensate enough for the free expansion due to stress relief in the polymer melt. Hence, the final size of the extrudate is still greater than the nozzle orifice size, even though the material has cooled down after extrusion. Die-swell is the response of the polymeric material subjected to intense pressure under a thermal environment, leading to dimensional changes in the extrudate after the extrusion. ABS, which stands for *acrylonitrile butadiene styrene*, is a commonly employed amorphous polymer of thermoplastic nature and is often used in *material extrusion* (MEX)-based AM processes. It is extruded at high temperatures and is not biodegradable in nature. It is regarded for its exceptional hardness and impact resistance. Computers, pipes, boats, and LEGO toys all use it (Algarni & Ghazali, 2021). Some of the other conventional thermoplastics used in AM are PA6 (*nylon*), *polyether ether ketone* (PEEK), *polyethylene terephthalate* (PET), *poly lactic acid* (PLA), and *thermoplastic polyurethane* (TPU) (Bakir et al., 2021). The die-swell behaviour is subjected to various processing conditions and material choice, and hence, should be meticulously studied, in order to achieve the optimum extrusion dimensions and form stability. Other research studies revolving around the progress in advanced manufacturing processes are Sharma et al. (2022); Sharma and Jain (2020); Sharma et al. (2018); Sharma et al. (2019a); Sharma et al. (2019b); Sharma et al. (2019c); Sharma et al. (2021a); Sharma et al. (2021b); Sharma et al. (2020); Sharma et al. (2022); Kalman et al. (2017); Lumay et al. (2019); Tripathi and Mallick (2017); Neveu et al. (2020a); Neveu et al. (2020b); Ratsimba et al. (2021); Tripathi et al. (2020); Tripathi et al. (2019); Tripathi et al. (2016); and Tripathi and Mallick (2014).

In the current chapter, various processing parameters in a screw extrusion process (such as peak temperature, axial thermal gradient, and extruder speed) are studied for the extrudate dimensions and the associated die-swell behaviour. A Taguchi's *design-of-experiments* (DoE) based optimization study is also conducted, to simultaneously optimize the process parameters considered the output flow rate and the extrusion dimensions (Mittal et al. 2022). Also, the presence of the nozzle has to be considered as it originates as an opening point for melt flow and, therefore, can change the dynamics of the flow. Not only the nozzle size, in terms of the orifice diameter (d_N), but also the nozzle length (l_N) and their ratio (l_N/d_N) affects the shear behaviour, melt pressure, and the corresponding extrusion size, i.e., die-swell.

5.1   MATERIALS AND METHODS

5.2.1   THERMOPLASTIC ABS

ABS thermoplastic is investigated and characterized in this study. It is a thermoplastic polymer that is commonly used in polymer processing and prototyping (FDM 3D printing). It is an amorphous polymeric material with a *glass transition temperature* (GTT) of about 105°C. ABS remains in a hard and relatively brittle "glassy" state below the GTT but starts flowing beyond that. ABS have three monomers: *Acrylonitrile*, 1,3-*Butadiene* and *Styrene*, as shown in Figure 5.2.

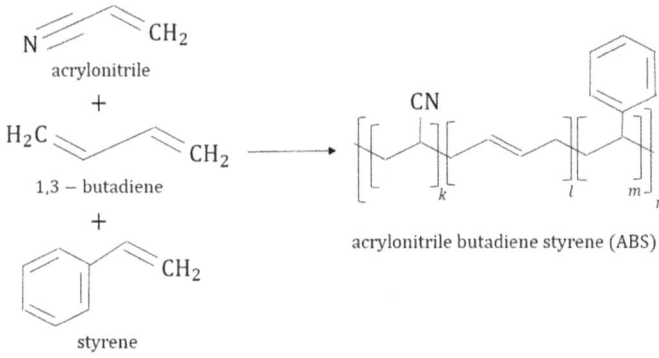

FIGURE 5.2   ABS and its constituent monomers.

ABS is modelled as a highly viscous thermoplastic using the Cross-WLF (*Cross-William–Landel–Ferry*) model, constituting an empirical formula connected with the principle of time-temperature superposition. It is used for polymer melts that have a high GTT. The model also takes care of process-related properties, other than temperature, like shear strain. The polymer melt viscosity, μ , of the solid/liquid polymer mixture is approximated by the Cross-WLF model in Equation 5.3 and 5.4, assuming that the polymer entering the channel is entirely compacted.

$$\mu = \frac{\mu_o}{1 + \left(\dfrac{\mu_o \cdot \dot{\gamma}}{\zeta^*} \right)^{1-n}} . \tag{5.3}$$

$$\mu_o = D_1 . exp \left[-\frac{A_1 . (T - T_a)}{A_2 + (T - T_a)} \right] \tag{5.4}$$

where μ_o is the viscosity at zero-shear and $\dot{\gamma}$ is the shear rate. T_a = GTT of the material. $\zeta^* = 29$ kPa, $n = 0.33$, $D_1 = 3.6.31 \times 10^{11}$ Pa-s, $A_1 = 27.21$ and $A_2 = 92.85$ K, are other material-related properties. The variation of ABS viscosity as predicted by the Cross-WLF model, with different shear rates for different temperatures, are shown in Figure 5.3.

5.2.2   SCREW EXTRUSION SETUP

A single-flight, constant pitch, single-screw extruder is fabricated along with all the associated parts. The conventional extruder screws, used in *injection moulding* industries, have three discrete zones with different pitch values and diameters (Alfaro et al., 2015). The distinctive change in the screw pitch helps in the pressure development but also loads up the drive motor. Additionally, the screw extruder for additive manufacturing purposes needs to be repeatedly started and stopped. This can generate a lot of stress on the drive motor. Therefore, a novel screw design

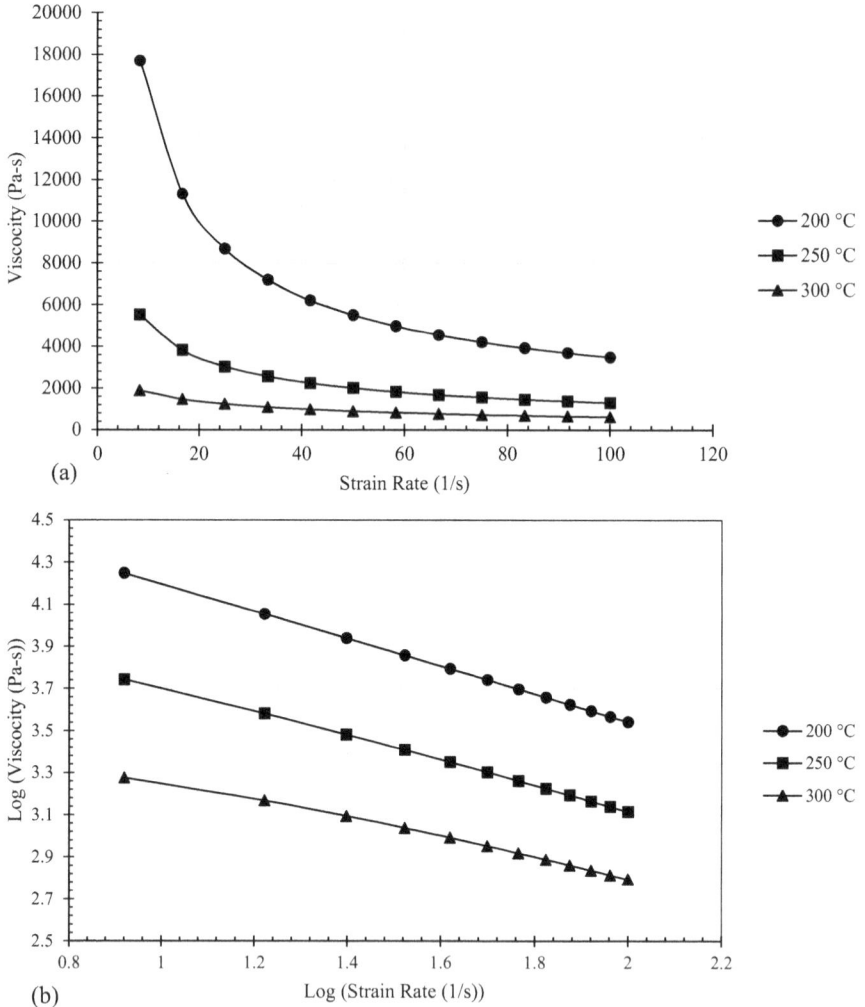

FIGURE 5.3 Variation of ABS melt viscosity with strain rate at different temperatures: (a) Linear scale, (b) Logarithmic scale.

is chosen, specially designed for material extrusion-based *additive manufacturing* applications. This is done, keeping in mind, that the primary function of the extruder screw, in this case, is to feed the polymer pellets into the heating zone and extruder the polymer melt through the nozzle. Therefore, a constant pitch extruder geometry is finalized, with a large helix angle and associated width and pitch. The screw is specifically designed for material conveying and is somewhere in between a conventional *injection moulding* screw and an auger drill design.

The helix angle is the single-most important design factor for the screw extruder. The screw geometry is meticulously designed for maximum flow rate by setting the

TABLE 5.1

Various Dimensions of the Extruder Screw Setup

Dimensions	Value
Screw Length, L	150 mm
Screw Diameter, D	16 mm
Helix Angle, $\varnothing$	20.84°
Channel Depth, H	4 mm
Metering Zone Length, L_m	67 mm
Nozzle Length, L_n	10 mm
Nozzle Diameter, d_n	1 mm

helix angle to the optimum value (Mittal et al., 2022). The optimized helix angle and the associated screw geometry parameters are mentioned in Table 5.1. The torque requirement and the generated stress are heavily reduced with this approach. A similar technique is also employed by Whyman et al. (2018), for the extrusion of biopolymer pellets for 3D printing. Kumar et al. (2018) also developed a comparable single-screw extruder setup, mounted on a 3-axis CNC for the fabrication of flexible objects.

The screw geometry profile and the fabricated screw–barrel–nozzle assembly are shown in Figure 5.4 (a & b). A small clearance of 0.05 mm is given between the screw and the barrel to minimize the leakage flow over the flights. This allows the screw to rotate uninterruptedly while any leakage flow can be neglected. The extruder screw and its housing barrel are plastic-friendly EN41B alloy steel. A bronze nozzle of 1 mm orifice diameter and 10 mm orifice length is thread attached to the barrel, giving the extrudate its final shape. Standard-sized thermoplastic pellets (~3.5 mm) are fed into the extrusion system via a hopper made of *stainless steel* (SS), with a capacity of 600 cm³. The heating system consists of a single resistance-based band heater placed circumferentially onto the barrel, which provides a uniform temperature. Separate contact-based temperature measurement via a J-type thermocouple and a PID-based feedback temperature control is also present. Aluminium-based circular fins for passive cooling are provided between the hopper and heating unit to avoid choking the throat area. The final assembly is shown in Figure 5.4 (c). High torque NEMA 34 stepper motor, with a torque capacity of 85 kg-cm, is used to drive the extruder screw via a 24 V DC power supply and a 5 Amp stepper motor driver. The motor speed (RPM) is controlled via the Arduino Nano microcontroller.

5.2.3 PROCESS MODELLING

A flow model is prepared, in terms of the output flow rate. The polymer material travels in a spiral trajectory along the length of the extruder. This spiralling volume is created by a rectangular segment with a thickness equal to the flight height, H, and

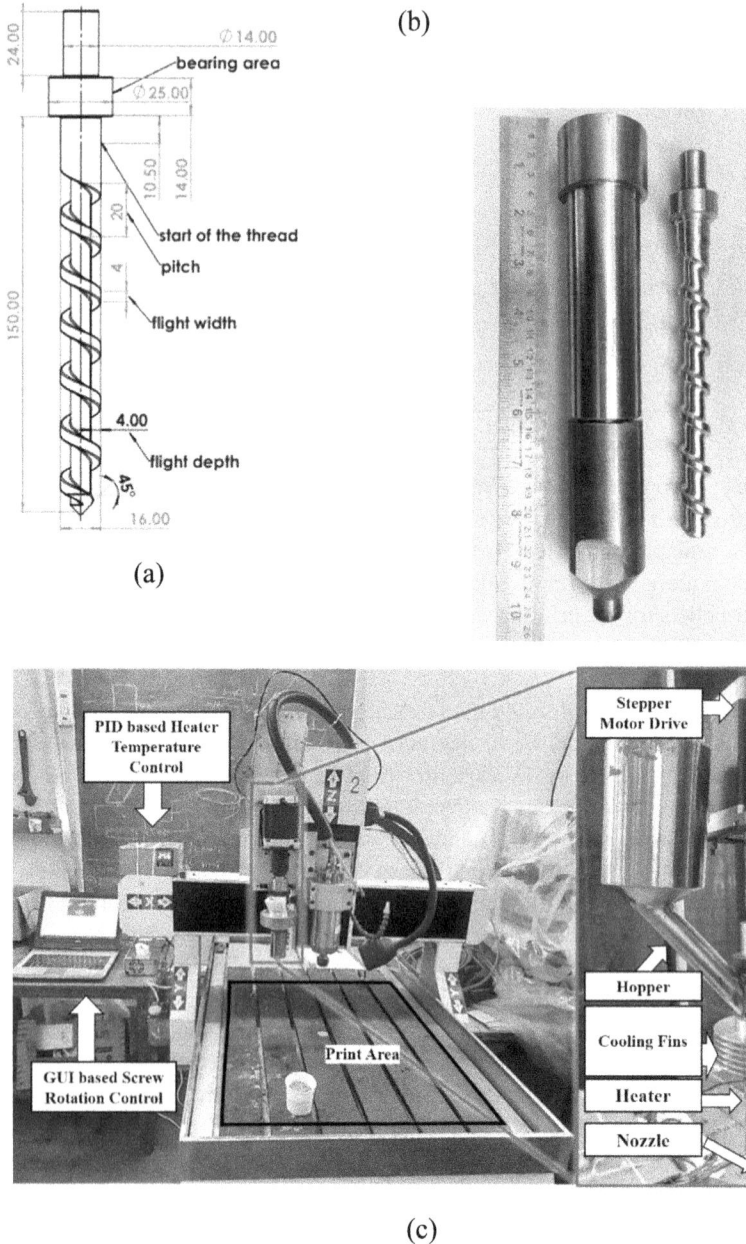

(b)

(a)

(c)

FIGURE 5.4 Extruder screw setup: (a) Screw geometry profile, (b) Fabricated parts, (c) Final Assembly.

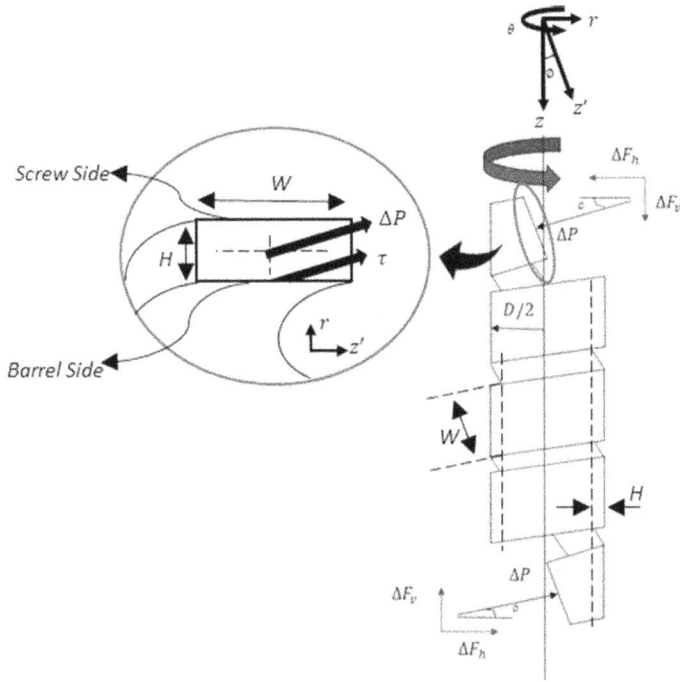

FIGURE 5.5 Schematics of the screw extrusion process.

a width corresponding to the flight separation, W (Figure 5.5). Several assumptions are established to govern the flow:

- The flow remains in a steady-state condition and exhibits Newtonian behaviour.
- The flight width exceeds the flight height ($W > H$).
- We assume that there are no leaks in the clearances.
- The polymer is entirely in a liquid state within the metering zone.

The screw turning induces the drag flow, which symbolizes the forward melt flow, whereas the reverse flow is generated by die pressure. The overall volumetric flow, Q, in Equation 5.5, combines the flow resulting from drag, Q_d, and the flow driven by pressure, Q_p, (Francis et al., 2016).

$$Q = Q_d + Q_p = \frac{1}{4}\pi^2.D^2.H.N.\sin(2\varnothing) - \frac{\Delta P.H^3.W}{12.\mu_M.L_M} \tag{5.5}$$

N represents the screw speed, ΔP stands for the pressure within the extruder screw, μ_M represents the melt viscosity in the metering zone, and L_M signifies the length of the metering zone. The pressure forces exert an opposing influence on the flow. Melt pressure serves as a crucial parameter in modelling and significantly impacts polymer flow. This model is akin to the widely employed generalized Newtonian method for contemporary screw designs with deeper channels. A similar outcome was derived by Crawford and Martin (2020), considering leakage flow. Nevertheless, for most practical applications, it suffices to disregard the leakage flow term. The viscosity of molten plastic reaches saturation, approximately 1000 Pa-s, under various processing conditions. Similar assumptions have been made by other researchers in characterizing the screw extrusion of thermoplastics (Feuerbach & Thommes, 2021).

The total volumetric flow rate, denoted as Q, is inherent to the screw extruder for a given temperature and screw speed. This same flow rate can be employed to calculate the velocity at the nozzle. The flow through the nozzle is determined using the Hagen-Poiseuille relation (Equation 5.6):

$$Q = A_N.v_N = \frac{\pi}{4}.d_N^2.v_N = \frac{\Delta P.\pi.d_N^4}{128.\mu_N.L_N} \tag{5.6}$$

Here, A_N represents the nozzle area, which corresponds to the nozzle diameter, denoted as d_N. Additionally, v_N, L_N, and μ_N represent the nozzle-extrusion velocity, nozzle length, and melt viscosity at the nozzle, respectively. Various dimensions of the nozzle are shown in Figure 5.6.

The final expression for the nozzle extrusion velocity (Equation 5.7) is calculated using Equations 5.5 and 5.6.

$$v_N = \cfrac{\pi.\left(\frac{D}{d_N}\right)^2.H.\sin(2\varnothing).N}{\left(1+\frac{32}{3}\left(\frac{L_N}{L_M}\right).\left(\frac{H^3.D}{d_N^4}\right).\sin^2\varnothing.\exp\left(\frac{A_1.(T_M-T_a)}{A_2+(T_M-T_a)}\frac{A_1.(T_N-T_a)}{A_2+(T_N-T_a)}\right)\cfrac{\left(1+\left(D_1.\exp\left(\frac{A_1.(T_M-T_a)}{A_2+(T_M-T_a)}\right).\frac{\pi.D.N.\cos(\varnothing)}{\zeta^*.H}\right)^{1-n}\right)}{\left(1+\left(D_1.\exp\left(\frac{A_1.(T_N-T_a)}{A_2+(T_N-T_a)}\right).\frac{8.v_N}{\zeta^*.d_N}\right)^{1-n}\right)}\right)} \tag{5.7}$$

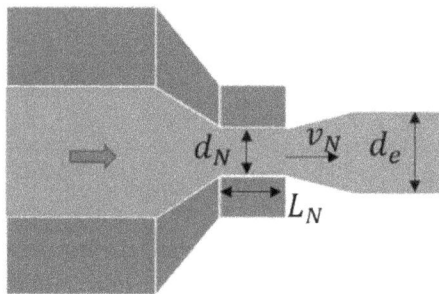

FIGURE 5.6 Graphical representation of flow through the nozzle.

The mean shear stress is calculated at the average shear rate and temperature (Fenner,1977). The strain rates in the nozzle ($\dot{\gamma}_N$) and the metering zone ($\dot{\gamma}_M$) are substituted with their corresponding physical parameters ($\dot{\gamma}_N = 8.v_N \, / \, d_N, \dot{\gamma}_M = \pi.D.N.\cos(\varnothing) \, / \, H$). Additionally, the extrusion velocity can be employed to ascertain critical process attributes, like total volumetric flow, rated torque, power input, and specific energy consumption (Mittal et al., 2023).

5.2.4 Process Parameters

As mentioned before, the screw extrusion process primarily has two process parameters defining and characterizing the process. These process parameters are the barrel temperature (T) and the rotational speed (N). Apart from these two, the die-swell behaviour is particularly affected by the nozzle dimensions as well, namely nozzle diameter (d_N) and nozzle length (l_N). The presented extruder design has a single heating coil that covers the whole nozzle and the metering section of the screw, such that they share almost the same temperature at steady-state conditions. Hence, the nozzle temperature (T_N) and the barrel temperature can be treated as the same, i.e. $T_N = T$. Therefore, for this research, four process parameters, namely, nozzle temperature, extruder's rotational speed, nozzle diameter, and nozzle length are investigated. It should be noted that instead as a single independent parameter, the nozzle length is studied as a ratio of the nozzle diameter, i.e. ($l_N \, / \, d_N$).

5.2.4.1 Nozzle Temperature (T_N)

The nozzle temperature, during the extrusion, is taken as a process parameter. It is also the extrusion temperature, i.e., the temperature at which the extrusion occurs. The current extruder setup design consists of a single resistive heater, placed coaxially around the barrel-nozzle assembly, covering the complete barrel and the metering zone of the extruder screw (Figure 5.4 (c)). The heater is controlled via a PID setup which provides a single uniform temperature at the steady state. The nozzle temperature directly affects the polymer viscosity and thus, defines the melt flow. For better fluidity, and high output, the nozzle temperature should be kept high, as for a given strain rate (screw speed), the polymer viscosity decreases with increasing nozzle temperature (Figure 5.3). It should also be noted that increasing the nozzle temperature beyond a certain limit may also have negative effects on the extrusion process. The polymer melt may degrade (burn-out) or cross-link, which can cause defects in the extrudate and reduce the output flow rate. For thermoplastic ABS, the extrusion temperature lies in the range of 215°C to 250°C (Cano-Vicent et al., 2021). There has not been any extensive research done on the die-swell behaviour of thermoplastic ABS beyond 250°C. Therefore, in this research, the extrusion temperature is varied from 250°C to 300°C, in the increments of 25°C.

5.2.4.2 Screw Speed (N)

The screw speed has a direct implication on the flow of the thermoplastic melt and the extrusion capabilities. From Equation 5.7, it can be seen that a higher speed results in escalated outputs. On the other hand, better flow metering is achieved at lower speeds. The conflict between productivity and flow control through metering leads to an optimization problem. For screw extrusion of thermoplastics, screw speed can be as high as 100 RPM or beyond. In this research, three discrete levels of screw speed are explored (70 RPM, 85 RPM, 100 RPM).

5.2.4.3 Nozzle Diameter (d_N)

The presence of the nozzle has to be considered as it originates as an opening point for melt flow and therefore, can change the dynamics of the flow. A smaller nozzle (lower diameter) not only reduces the output flow rate but also tremendously increases the melt pressure inside. There is a 16-factor increase in the melt pressure just by reducing the nozzle size by half (Equation 5.5). This makes smaller nozzle more susceptible to die-swell and extrudate expansion after free extrusion. Although, a smaller nozzle is preferable for better metering and precise deposition. A bigger nozzle (larger diameter) on the other hand, delivers a greater output and significantly reduces the melt pressure but does not provide proper metering and is difficult to control the flow for precision deposition. There is always a compromise between the nozzle size and the corresponding melt pressure generation and flow metering. The commonly used nozzle size is 0.4 mm (Chacon et al., 2021) but can go even up to 0.2 mm on the smaller end and 1 mm on the larger end. Custom nozzle, larger than 1 mm can also be made, as per requirements. For the presented research, four different (and custom) nozzle sizes are studied, starting from 0.2 mm, 0.5 mm, 1 mm to 2.5 mm.

5.2.4.4 Nozzle Length (l_N)

The length of the nozzle has an effect on the die-swell behaviour, during screw extrusion of thermoplastics. As seen from Equation 1.5, the nozzle length has a direct relation to the polymer melt pressure inside the nozzle. The effect of nozzle length is usually studied together with the nozzle diameter, as a dimensionless ratio (l_N/d_N). Hence, the same nozzle length can be used over multiple nozzle sizes (as it is convenient to manufacture variable diameter rather than variable length), to get different l_N/d_N values. These ratios can vary from 5 to even 50, or beyond, depending upon the given geometry of the nozzle. In general, a longer nozzle length can result in a greater degree of die-swell due to increased residence time and shear stress on the material within the nozzle. This can be attributed to the fact that as the molten polymer flows through the nozzle, it experiences a higher level of shear stress and undergoes greater orientation, leading to a higher degree of orientation and alignment of polymer chains. This, in turn, can result in increased die-swell as the oriented chains attempt to relax back to their original random orientation upon exiting the die. However, a considerably long nozzle (high l_N/d_N) can result in an even higher degree of orientation such that the extrusion state is locked, i.e., the size of the extrudate conforms to the size of the nozzle and hence, experiences low die-swell (Liang, 2008). Not only the nozzle length, but the l_N/d_N value over which the extrusion is happening, dictates the die-swell behaviour.

However, the effect of nozzle length on die-swell behaviour is also influenced by other factors, such as the melt viscosity of the material, the geometry of the die, and the processing conditions used. Therefore, it is essential to consider all of these factors together to optimize the extrusion process and minimize die-swell behaviour. In the reported work, four different nozzles of four different sizes (diameter) but the same nozzle length (10 mm) are used. Hence, four discrete values of l_N/d_N are investigated.

5.3   RESULTS AND DISCUSSIONS

This section describes the results obtained from the physical experiments. The results describe the effect of various process parameters and their derivatives (like nozzle pressure and shear stress) on the die-swell characteristics of screw-extruded thermoplastic ABS.

5.3.1   EXPERIMENTAL RESULTS

Die-swell is a measure of extrusion characteristics and depends heavily on the exit conditions of the material after being extruded from the nozzle. The single-most important exit parameter for material extrusion is the extrusion flow rate which is often given in terms of the extrusion velocity at the nozzle, v_N. An analytical model is presented (in section 5.2.3) which closely captures the nozzle extrusion velocity (Equation 5.7) for a given set of process parameters (N, T, d_N, l_N). To physically verify the presented model for nozzle extrusion velocity, experimental validation is done. The model is verified for a particular nozzle set, with $d_N = 1$ mm but can be done for any nozzle size in general. It should be noted that only the process parameters of screw speed and barrel temperature are varied, in real-time, while the nozzle length and diameter are set to 10 mm and 1 mm, respectively. The extrusion velocity is determined, in "mm/min" by measuring the length of the extrusion, l_E in "mm" for 10 seconds followed by necessary unit conversion, as shown in Equation 5.8.

$$v_N = \frac{l_E}{(10/60)}\, mm\,/\,min \qquad (5.8)$$

The system is loaded with thermoplastic ABS pellets through the hopper entrance. In Figure 5.7(a), the physical extrusion process is depicted, including various nozzle geometries. Commercially available ABS pellets are used with an average size of 3.5 mm, as shown in Figure 5.7(b). The data from experiments, along with the corresponding process parameters, are collated in Table 5.2. Each experiment is repeated ten times, and only the average value is reported, as the standard deviation consistently remains below 10% of the mean value in all cases. A strong correlation is evident between the experimental and analytical results. Figure 5.7(c) displays the deviation from linearity, with a maximum value of only 15.72%.

5.3.1.1   Analysis of Correlation

In order to assess the effectiveness of the model, a correlation analysis is conducted, comparing the experimental data to the analytically predicted values. The outcomes of the experiments are treated as independent variables, denoted as x, while the

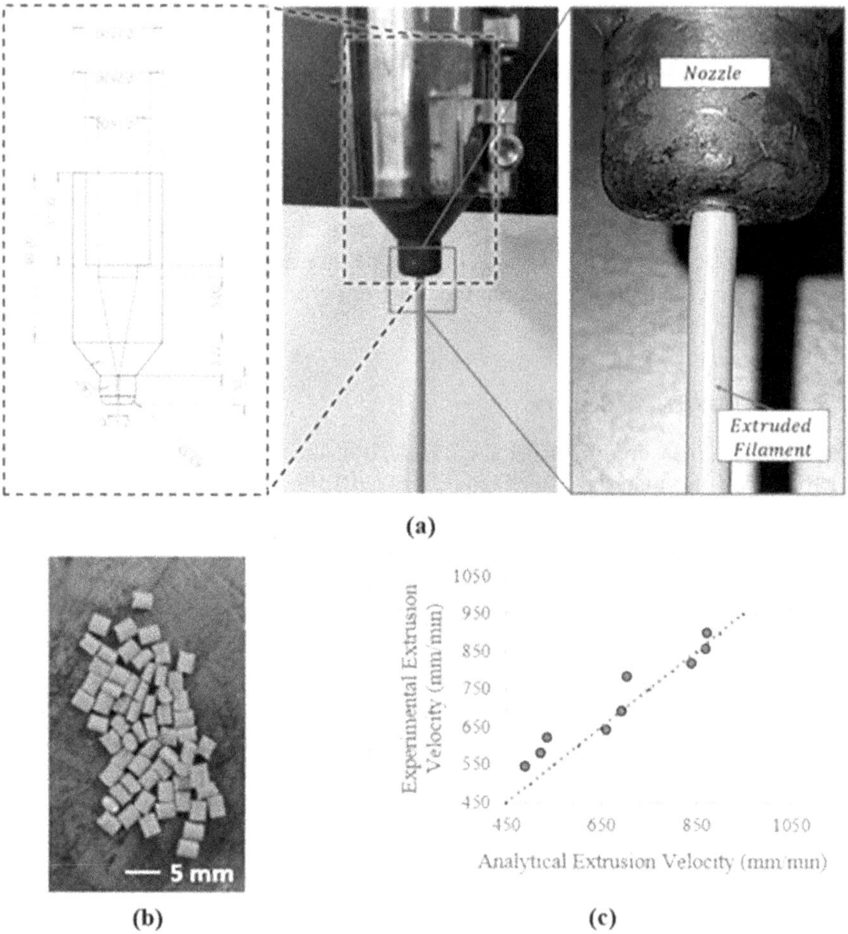

(a)

(b)

(c)

FIGURE 5.7 (a) Physical extrusion process with a magnified view of the nozzle area and nozzle specifications, (b) Commercially available ABS pellets, employed in this study, (c) Discrepancies between the analytical and experimental findings indicating a slight deviation in the results.

analytical results obtained from the model serve as the dependent variable, denoted as y. Equation 5.9 is applied to calculate the correlation coefficient r.

$$r = \left(\frac{1}{n-1} \right) \cdot \Sigma \left(\frac{x - \mu_x}{\sigma_x} \right) \cdot \left(\frac{y - \mu_y}{\sigma_y} \right) \tag{5.9}$$

The variable n represents the number of x–y pairs, and μ and σ denote their respective mean and standard deviation. The obtained correlation coefficient is 0.96568, indicating a strong correlation of 96.57%.

TABLE 5.2
Analytical and Experimental Extrusion Velocity for Various Process Parameters

Sr. No.	Screw Speed, N (in RPM)	Barrel Temperature, T (in °C)	Nozzle Extrusion Velocity, v_N (in mm/min)	
			Theoretical	Physical
	40	230	536.34	620.68
	40	240	521.24	580.64
	40	250	490.16	545.45
	50	230	703.25	782.60
	50	240	693.21	692.30
	50	250	661.33	642.85
	60	230	873.90	900.12
	60	240	870.80	857.14
	60	250	840.29	818.18

5.3.1.2 Curve Fitting

The final equation describing the extrusion velocity (Equation 5.7) is intricate and relies on various geometrical and process parameters. To simplify the determination of extrusion velocity (in mm/min), a closed-form solution is established using MATLAB through a non-linear least square optimization approach. In a setup involving a screw extruder, the readily adjustable parameters are the barrel temperature T (in °C) and screw rotational speed N (in RPM). Equation 5.10 represents the empirical relation derived from this analysis, which can be directly applied to estimate nozzle extrusion velocity and related parameters such as volumetric flow and mass throughput. This estimation is feasible for any given liquefier temperature and screw speed within the operational range specified for the described setup and the aforementioned thermoplastic material. This study's findings provide a basis for further investigations into process optimization and operational control.

$$v_N(N,T) = 6131 + 3.385 \times N - 43.89 \times T + 0.1445 \times N^2 - 0.01677 \times N \times T + 0.08287 \times T^2$$

(5.10)

5.3.2 EFFECT OF PROCESS PARAMETERS

5.3.2.1 Nozzle Temperature

The variation of die-swell along with the nozzle temperature at different nozzle shear stress is shown in Figure 5.8 (a). From the graph, it can be seen that the extent of die-swell decreasing with increasing nozzle temperature. This can be directly inferred from the temperature dependence of the polymer melt viscosity. The polymer melt viscosity has an inverse relation with the melt temperature (Figure 5.3).

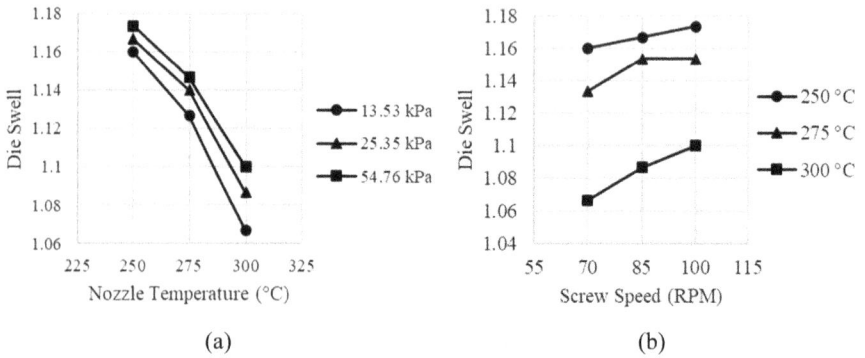

FIGURE 5.8 (a) Variation of die-swell with nozzle temperature at different nozzle shear stress, (b) Variation of die-swell with screw speed at different nozzle temperatures.

As the nozzle temperature increases, the melt temperature also increases, which in turn reduces the melt viscosity. The reduction in melt viscosity reduces the pressure built-up in the nozzle and the corresponding state-of-stress is also reduced. The polymer melt is more relaxed/stress-free at higher temperatures, which is reflected in a low die-swell after extrusion.

A counter effect is observed for die-swell with the induced nozzle shear stress, i.e., as the nozzle shear stress increases, the die-swell ratio increases. The nozzle shear has a direct dependence on the shear strain which in turn has a direct correlation with the screw speed (N). Hence, it can also be said that the die-swell increases with increasing screw speed.

5.3.2.2   Screw Speed

As mentioned before, the die-swell has a direct dependence on the screw speed, i.e., the die-swell increases with the increasing screw speed. The trend is almost linear between the screw speed and the die-swell. The intercept of this N vs B curve (Figure 5.8 (b)) is dependent on the nozzle temperature. The die-swell reduces with the increasing nozzle temperature, for a given speed. Hence, the intercept reduces with increasing nozzle temperature.

5.3.2.3   Nozzle Diameter

The effect of the nozzle size (diameter) on the die-swell characteristics of screw extruded thermoplastic ABS, at various screw speeds, is shown in Figure 5.9 (a). From the results, it is observed that the die-swell tendency increases with increasing nozzle size. The result seems counter-intuitive, because the nozzle diameter has an inverse bi-quadratic relation with the generated nozzle pressure (Equation 5.6).

A different approach is to look at the effective area for heat transfer in the nozzle, between the polymer melt and the nozzle inner surfaces. As the nozzle diameter increases, the effective area of radial heat transfer ($\pi.d_N.l_N$) also proportionally increases. This increment in the effective heating leads to a higher degree of melting and melt temperature, which increases the thermal strain inside the polymer. This

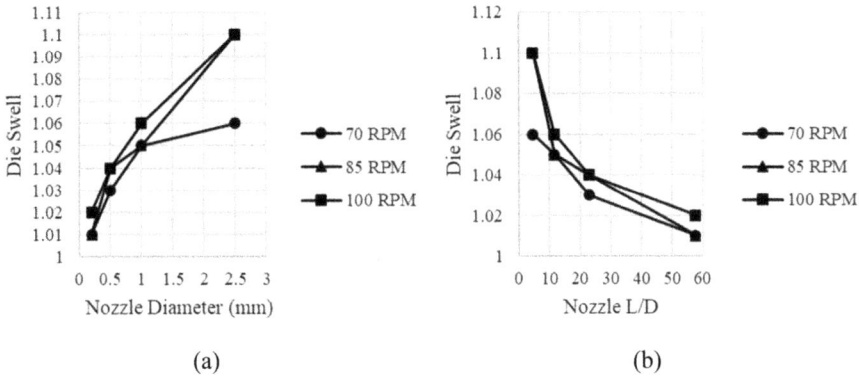

(a)   (b)

FIGURE 5.9 (a) Variation of die-swell with nozzle diameter at different screw speeds, (b) Variation of die-swell with nozzle lengths at different screw speeds.

increased thermal strain can lead to an increased state-of-stress in the polymer melt which is reflected as an increased die-swell after the extrusion.

It is also observed that screw speed doesn't have much influence on the die-swell behaviour over various nozzle sizes. Therefore, nozzle diameter precedes the screw speed in die-swell.

5.3.2.4 Nozzle Length

The effect of nozzle length is captured as its ratio with the nozzle diameter (l_N/d_N). The results show that the die-swell reduces with the increasing nozzle length, as shown in Figure 5.9 (b). This again seems counter-intuitive from the nozzle pressure equation (Equation 5.6). But, similar to the nozzle diameter, the inverse effect of nozzle length on the die-swell can be explained by the heating enhancement analogy. As the nozzle length increases, the effective area of radial heat transfer ($\pi.d_N.l_N$) also increases, which in turn elevates the melt temperature. This reduced the melt viscosity in the nozzle (Figure 5.3), which results in a reduced melt pressure. But the nozzle pressure also proportionally increases, with the increase in nozzle length. This generates a countering action to the state-of-stress, from the perspective of melt pressure. From the results given below, it can be said that the effect of increased temperature on lowering the melt pressure via reduction in the melt viscosity in the nozzle, is greater than the direct increase in the melt pressure by the nozzle length. Hence, the overall melt pressure reduces, which lowers the state-of-stress in the extruded polymer and reduces the die-swell. Similar to the previous case, a significant effect of screw speed on the die-swell behaviour for a given nozzle length, is not observed.

From the above discussions, it can be stated that the geometrical aspects of the nozzle heavily influence the die-swell behaviour and even precede the processing parameter of screw speed. The die-swell behaviour is also investigated with respect to other quantities such as nozzle strain rate and nozzle shear stress.

The die-swell ratio of the screw extruded ABS thermoplastic is investigated over the average of the maximum strain rate in the nozzle, for a given screw speed, over

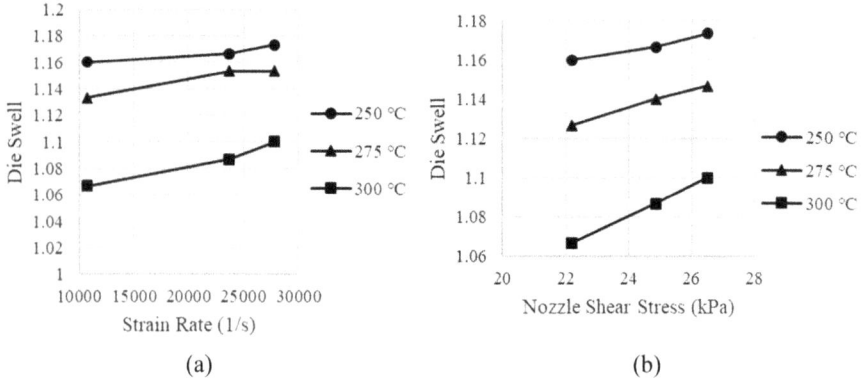

(a)

(b)

FIGURE 5.10   (a) Die-swell vs nozzle strain rate, (b) Die-swell vs nozzle shear stress.

various nozzle temperatures (Figure 5.10 (a)). The maximum nozzle strain rate ($\dot{\gamma}_N$) is calculated using the Newtonian-based parabolic velocity profile, at the nozzle boundary, i.e., $r = R_N$, as shown in Equation 5.11.

$$\left.\dot{\gamma}_N\right|_{max} = \left.\frac{dv_N}{dr}\right|_{r=R_N} = \left.\frac{d}{dr}\left(\frac{\Delta P}{4.\mu_N.l_N}.\left(R_N^2 - r^2\right)\right)\right|_{r=R_N} = \frac{\Delta P.R_N}{2.\mu_N.l_N} \tag{5.11}$$

The average of this maximum strain rate is calculated over the different screw speeds (70, 85, 100 RPM) for a given nozzle temperature. It is observed that the die-swell increases with the strain rate, which is expected as it shifts the polymer melt towards a greater state-of-stress.

The nozzle shear stress (τ) can also be calculated using the shear rates (Equation 5.12). As it holds a direct relation with the nozzle shear, the die-swell also increases with the increasing shear stress (Figure 5.10 (b)). The temperature has a declining effect on the die-swell for a given shear stress. Again, this can be attributed to the thermal relaxation in the polymer melt with increasing temperature

$$\tau = \mu_N.\dot{\gamma}_N \tag{5.12}$$

5.3.3   PROCESS OPTIMIZATION

Die-swell is the material's response towards stress relaxation after the extrusion. It often leads to an increased extrudate size, as compared to the nozzle diameter, and, hence, should be minimized for better dimensional conformance. In this section, optimization of a screw extrusion process is investigated with the objective of minimal die-swell while maintaining maximum extrusion rate (nozzle velocity), for a given nozzle size of 2.5 mm.

Taguchi's method-based process optimization is done using L_9 (3^3) orthogonal array arrangement for screw extrusion of thermoplastic ABS. Process parameters such as screw speed, peak temperature (nozzle temperature), and thermal gradient are simultaneously optimized for high output and dimensional stability by maximizing the specific velocity of extrusion, which is defined as the ratio of extrusion velocity to the extrudate size.

5.3.3.1 Specific Velocity (v_s)

The extrusion velocity (v_N) is heavily dependent upon process parameters and directly defines the flow output. Another parameter that directly defines the flow output is the extruded filament size (d), which is also, in turn, dependent upon process parameters. Because of their dependence on similar quantities, extrusion velocity and extrudate diameter are studied together in terms of specific velocity, v_s which is defined in Equation 5.13.

$$v_s(1/min) = \frac{v_N(mm/min)}{d(mm)} \tag{5.13}$$

5.3.3.2 Thermal Gradient (T_x)

Apart from the screw speed and nozzle temperature, the thermal gradient (°C/m) along the length of the extruder is taken as a design factor. Thermal gradient is an important parameter that captures the essence of the spread of temperature over the length of the system. Along with peak temperature, the temperature gradient defines the overall thermal profile. Together they define the temperature spread over the screw-barrel system, which directly affects the melt viscosity and subsequently extrusion characteristics.

5.3.3.3 Design of Experiments (DoE)

Taguchi's method-based *design of experiments* trial is conducted to optimize the process parameters. Three factors are identified as the process parameters, i.e., screw speed, peak temperature, and thermal gradient. Corresponding to each factor, three levels are identified based on commonly used values (Algarni & Ghazali, 2021; Altinkaynak et al., 2011), as shown in Table 5.3. Hence, an L_9 orthogonal array for a 3×3 arrangement is used. Repeatability is also required for process validation and optimization.

TABLE 5.3
Various Factors and Their Associated Levels

Factors	Levels		
	1	2	3
Peak Temperature (°C)	250	275	300
Temperature Gradient (°C/m)	250	350	450
Screw Speed (RPM)	70	85	100

5.3.3.4 DoE Results

Physical experimentation is done using the indigenously fabricated single-screw extruder setup, where the system is tuned to the processing conditions, as suggested by the DoE. Parameters such as extrusion velocity and extrudate size are measured for each experiment with a repetition of ten times, and the average value is mentioned in Table 5.4. The mean of the means is determined for each factor and corresponding level. *Signal-to-Noise* ratio (S/N) is also determined, using Equation 5.14, where (*n*) is the number of observations and (*y*) is the observed response; v_s, in this case. Both the values are set to "larger is better" (Li et al., 2019) for optimizing process parameters with maximum extrusion velocity and minimal die-swell. The results of the optimization process are shown in Figure 5.11.

$$\frac{S}{N} = -10\log\left(\frac{1}{n}\sum_{i=1}^{n}\frac{1}{y_i^2}\right) \tag{5.14}$$

TABLE 5.4

Experimental Results for Taguchi's Method-based L₉ Orthogonal Array

Experiment	Design Parameters			v_N	d	v_s
	T (°C)	T_x (°C/m)	N (RPM)	(mm/min)	(mm)	(1/min)
1	250	250	70	178	2.9	61.38
2	275	350	70	225.56	2.9	77.78
3	300	450	70	300	2.65	113.2
4	275	250	85	194.26	2.9	66.97
5	300	350	85	157.69	2.7	58.4
6	250	450	85	171.6	2.95	58.17
7	300	250	100	184.9	2.75	67.23
8	250	350	100	179.8	3	59.93
9	275	450	100	210.16	3	70.05

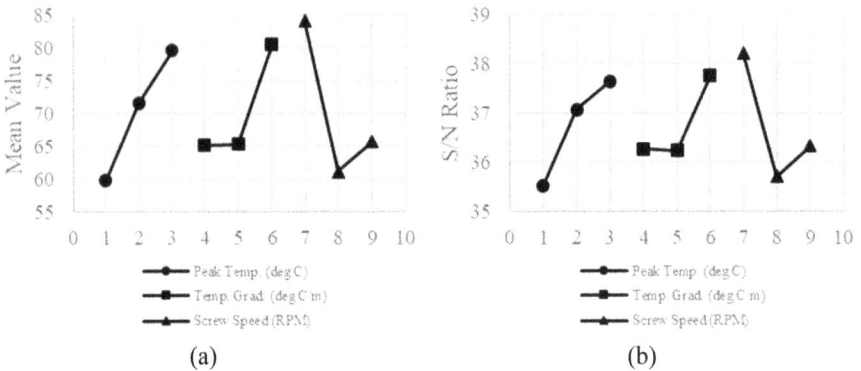

FIGURE 5.11 DoE Results, (a) Mean of the means, (b) S/N Ratios.

TABLE 5.5
Rank of the Design Parameters

Factors	Levels			Max.-Min.	Rank
	1	2	3		
Peak Temp. (°C)	59.82667	71.6	79.61	19.78333	2
Temp. Grad. (°C /m)	65.19333	65.37	80.47333	15.28	3
Screw Speed (RPM)	84.12	61.18	65.73667	22.94	1

It is observed that optimum working conditions occur at the highest peak temperature (300°C), highest thermal gradient (450°C/m), and lowest speed (70 RPM). This is because, at the highest workable peak temperature, the melt viscosity (and so the flow resistance) is minimal. The highest thermal gradient ensures that a necessary temperature difference is maintained. The lowest working screw speed provides the maximum time for thermal interaction for proper melt and metering.

The rank of each factor is determined (Table 5.5) to estimate the dominance of each factor. Table 5.5 shows that the screw speed has the most dominating effect, while the temperature gradient has the least dominating effect on output per unit filament size (i.e., specific velocity). Hence, the screw rotational speed followed by the peak temperature in the system should be meticulously monitored and controlled for the best working conditions.

The process of screw extrusion follows a complex and implicit relation with various geometric and process parameters. For optimum results, peak temperature and thermal gradient should be as high as possible. Moreover, screw speed has the direct and the most dominating effect on the extrusion capabilities, while the temperature gradient has the least. Based on the above results, it is sufficient enough to correlate the process output in terms of screw speed and peak temperature.

5.4 CONCLUSIONS

Thermoplastic ABS is a widely used polymer material, commonly processed via screw extrusion methods. Screw extrusion is a versatile and a high-pressure process that uses a rotating screw extruder to process the polymeric material. During the processing of ABS using screw extrusion, the material experiences a tremendous amount of stress that is instantly released after extrusion. As a result of stress-relaxation in the polymer melt, after the extrusion, the melt expands, resulting in a size greater than the nozzle diameter. This phenomenon of melt expansion after extrusion is called die-swell. It is the physical manifestation of the material after processing and should be taken into consideration as it leads to dimensional changes and affects the conformance to the required shape. This research has experimentally studied the consequences of various processing parameters on the die-swell behaviour of thermoplastic ABS, processed via an indigenously fabricated single-screw extrusion setup. The melt rheology of the thermoplastic is characterized by the Cross-WLF viscosity model. The extruder setup is custom built for high-deposition applications like additive manufacturing. A flow-based process modelling is conducted wherein four investigation parameters are identified: two process parameters (screw speed and nozzle temperature) and two geometrical parameters (nozzle diameter and length). Physical experimentations are done to measure the induced die-swell in the polymer melt after processing. The findings indicate that the screw speed and nozzle diameter have a positive effect on the die-swell, while the nozzle temperature and length show a decreasing behaviour. To ensure dimensional stability, die-swell, which can be as high as 1.1 times the nozzle diameter, should be minimized. These results can help manufacturers optimize their ABS processing parameters and improve the dimensional accuracy of their products.

REFERENCES

Alfaro, J. A., Grünschloß, E., Epple, S., & Bonten, C. (2015). Analysis of a single screw extruder with a grooved plasticating barrel–Part I: The melting model. *International Polymer Processing*, *30*(2), 284–296. 10.3139/217.3021.

Algarni, M., & Ghazali, S. (2021). Comparative study of the sensitivity of pla, abs, peek, and petg's mechanical properties to FDM printing process parameters. *Crystals*, *11*(8), 995. 10.3390/cryst11080995.

Altınkaynak, A., Gupta, M., Spalding, M. A., & Crabtree, S. L. (2011). Melting in a single screw extruder: Experiments and 3D finite element simulations. *International Polymer Processing*, *26*(2), 182–196. 10.3139/217.2419.

Bakır, A. A., Atik, R., & Özerinç, S. (2021). Mechanical properties of thermoplastic parts produced by fused deposition modeling: A review. *Rapid Prototyping Journal*, *27*(3), 537–561. 10.1108/RPJ-03-2020-0061.

Cano-Vicent, A., Tambuwala, M. M., Hassan, S. S., Barh, D., Aljabali, A. A., Birkett, M., Arjunan, A., & Serrano-Aroca, A. (2021). Fused deposition modelling: Current status, methodology, applications and future prospects. *Additive Manufacturing*, *47*, 102378. 10.1016/j.addma.2021.102378.

Chacon, J. M., Caminero, M. A., Nunez, P. J., García-Plaza, E., & Becar, J. P. (2021). Effect of nozzle diameter on mechanical and geometric performance of 3D printed carbon fibre-reinforced composites manufactured by fused filament fabrication. *Rapid Prototyping Journal*, *27*(4), 769–784. 10.1108/RPJ-10-2020-0250.

Crawford R. J., & Martin, P. (2020). *Plastics Engineering* (Fourth edition, pp. 280–292). Butterworth-Heinemann.

Das, M. (2008). Effect of screw speed and plasticizer on the torque requirement in single screw extrusion of starch-based plastics and their mechanical properties. *Indian Journal of Chemical Technology*, *15*, 555–559.

Fenner, R. T. (1977). Developments in the analysis of steady screw extrusion of polymers. *Polymer*, *18*(6), 617–635. 10.1016/0032-3861(77)90066-0.

Feuerbach, T., & Thommes, M. (2021). Design and characterization of a screw extrusion hot-end for fused deposition modeling. *Molecules*, *26*(3), 590. 10.3390/molecules26030590.

Francis, L. F., Stadler, B. J. H., & Roberts, C. C. (2016). *Materials Processing: A Unified Approach to Processing of Metals, Ceramics and Polymers*. Academic Press. 10.1016/C2009-0-64287-2.

Li, J., Cai, J., Zhong, L., Cheng, H., Wang, H., & Ma, Q. (2019). Adsorption of reactive red 136 onto chitosan/montmorillonite intercalated composite from aqueous solution. *Applied Clay Science*, *167*, 9–22. 10.1016/j.clay.2018.10.003.

Kalman, H., Tripathi, N. M., Gabrieli, O. G., & Portnikov, D. (2017). Phase diagrams for pneumatic and hydraulic conveying. In *18th International Conferences on Transport and Sedimentation of Solid Particles, T and S 2017* (pp. 145–152). Wydawnictwo Uniwersytetu Przyrodniczego we Wroclawiu.

Kumar, N., Jain, P. K., Tandon, P., & Pandey, P. M. (2018). Extrusion-based additive manufacturing process for producing flexible parts. *Journal of the Brazilian Society of Mechanical Sciences and Engineering, 40*, 1–12. 10.1007/s40430-018-1068-x.

Liang, J. Z. (2008). Effects of extrusion conditions on die-swell behavior of polypropylene/diatomite composite melts. *Polymer Testing, 27*(8), 936–940. 10.1016/j.polymertesting.2008.08.001.

Lumay, G., Tripathi, N. M., & Francqui, F. (2019). How to gain a full understanding of powder flow properties, and the benefits of doing so. *ONdrugDelivery, 2019*, 42–47.

Mittal, Y. G., Gote, G., Kamble, P., Hodgir, R., Mehta, A. K., & Karunakaran, K. P. (2022). Significance of specific velocity in screw extrusion. *Proceedings of the 4th International Conference on Advances in Manufacturing Technology.*

Mittal, Y. G., Kamble, P., Gote, G., Patil, Y., Mehta, A. K., & Karunakaran, K. P. (2023). A novel analytical model for screw extrusion of thermoplastic ABS with emphasis on additive manufacturing. *Manufacturing Letters, 35*, 652–657. 10.1016/j.mfglet.2023.08.054.

Mount, E. M. (2017). Extrusion processes. In *Applied Plastics Engineering Handbook* (pp. 217–264). William Andrew Publishing.

M Tripathi, N., & S Mallick, S. (2017). Pneumatic conveying of Fly Ash: Bend Models investigation. *Advanced Materials Proceedings, 2*(8), 526–531.

Neveu, A., Lumay, G., Pillitteri, S., Monsuur, F., Pauly, T., Ribeyre, Q., Francqui, F., Vandewalle, N., & Tripathi, N. M. (2020b). Physical characterization of blends containing mesoporous particles with a focus on electrostatic properties. In *2020 AIChE Spring Meeting and 16th Global Congress on Process Safety*. AIChE.

Neveu, A., Tripathi, N. M., Rigo, O., Francqui, F., & Lumay, G. (2020a). Experimental investigation of spreadability of metal powders in recoating process. In *Proceedings - Euro PM2020 Congress and Exhibition (Proceedings - Euro PM2020 Congress and Exhibition)*. European Powder Metallurgy Association (EPMA).

Olabisi, O., & Adewale, K. (2016). *Handbook of Thermoplastics* (Second edition). CRC Press.

Ratsimba, A., Zerrouki, A., Tessier-Doyen, N., Nait-Ali, B., André, D., Duport, P., Neveu, A., Tripathi, N., Francqui, F., & Delaizir, G. (2021). Densification behaviour and three-dimensional printing of Y2O3 ceramic powder by selective laser sintering. *Ceramics International, 47*(6), 7465–7474. https://doi.org/10.1016/j.ceramint.2020.11.087

Reddy, B. V., Reddy, N. V., & Ghosh, A. (2007). Fused deposition modelling using direct extrusion. *Virtual and Physical Prototyping, 2*(1), 51–60. 10.1080/17452750701336486.

Rosato, D. V., Rosato, D. V., & Rosato, M. (2004). *Plastic Product Material and Process Selection Handbook*. Elsevier.

Sharma, A., Babbar, A., Tian, Y., Pathri, B. P., Gupta, M., & Singh, R. (2022). Machining of ceramic materials: A state of the art review. *International Journal on Interactive Design and Manufacturing (IJIDeM)*, 1–21. https://doi.org/10.1007/s12008-022-01016-7.

Sharma, A., & Jain, V. (2020). Experimental investigation of cutting temperature during drilling of float glass specimen. *IOP Conference Series: Materials Science and Engineering* (Vol. 715, No. 1, p. 012050). IOP Publishing.

Sharma, A., Jain, V., & Gupta, D. (2018). Characterization of chipping and tool wear during drilling of float glass using rotary ultrasonic machining. *Measurement, 128*, 254–263.

Sharma, A., Jain, V., & Gupta, D. (2019a). Tool wear analysis while creating blind holes on float glass using conventional drilling: A multi-shaped tools study. In *Advances in Manufacturing Processes: Select Proceedings of ICEMMM 2018* (pp. 175–183). Springer Singapore.

Sharma, A., Jain, V., & Gupta, D. (2019b). Comparative analysis of chipping mechanics of float glass during rotary ultrasonic drilling and conventional drilling: For multi-shaped tools. *Machining Science and Technology*, 23(4), 547–568.

Sharma, A., Jain, V., & Gupta, D. (2019c). Multi-shaped tool wear study during rotary ultrasonic drilling and conventional drilling for amorphous solid. *Proceedings of the Institution of Mechanical Engineers, Part E: Journal of Process Mechanical Engineering*, 233(3), 551–560.

Sharma, A., Jain, V., & Gupta, D. (2021a). Effect of pre and post tempering on hole quality of float glass specimen: For rotary ultrasonic and conventional drilling. *Silicon*, 13, 2029–2039.

Sharma, A., Jain, V., Gupta, D., & Babbar, A. (2020). A review study on miniaturization. In *Advanced Manufacturing and Processing Technology* (First edition, pp. 111–131). CRC Press.

Sharma, A., Kalsia, M., Uppal, A. S., Babbar, A., & Dhawan, V. (2022). Machining of hard and brittle materials: A comprehensive review. *Materials Today: Proceedings*, 50, 1048–1052.

Tripathi, N. M., Francqui, F., & Lumay, G. (2020). Influence of relative air humidity on the flow property of fine powders. In *Third International Conference on Powder, Granule and Bulk Solids: Innovations and Applications PGBSIA 2020 February 26–28, 2020* (p. 63).

Tripathi, N. M., Francqui, F., Pirenne, T., & Lumay, G. (2019). Measuring food powders electrical properties as a result of anti-static content. In *2019 AIChE Annual Meeting*. American Institute of Chemical Engineers.

Tripathi, N. M., Levy, A., & Kalman, H. (2016). Initial acceleration pressure drop in dilute phase pneumatic conveying system. In *Powder, Granule and Bulk Solids: Innovations and Applications Conference*.

Tripathi, N. M., & Mallick, S. S. (2014). *An Investigation into Pressure Drop Across Bends for Fluidised Densephase Pneumatic Conveying Systems* (Doctoral dissertation).

Voon, S. L., An, J., Wong, G., Zhang, Y., & Chua, C. K. (2019). 3D food printing: A categorised review of inks and their development. *Virtual and Physical Prototyping*, 14(3), 203–218. 10.1080/17452759.2019.1603508.

Whyman, S., Arif, K. M., & Potgieter, J. (2018). Design and development of an extrusion system for 3D printing biopolymer pellets. *The International Journal of Advanced Manufacturing Technology*, 96, 3417–3428. 10.1007/s00170-018-1843-y.

6 A Prospective View of Biomaterials

3-D Printing of Biomedical Applications

Priyanka Yadav, Pratibha Tiwari,
Kusum Yadav, and Hariom Sharma

6.1 INTRODUCTION

Three-dimensional printing is seen as a significant advancement in material engineering, particularly in customising biomedicine. An effective bioink is thus a stepping stone to 3-dimensional printing techniques employed in the biomedicine sector. Silk fibroin has been used as a biomaterial for decades because of its remarkable higher machinability, better biocompatibility, and biodegradability, which make it as a possible choice of bio-ink for 3-dimensional printing (Wang et al., 2019).

Three-dimensional printing demonstrates the direct layer-by-layer manufacturing of components under the direction of digital data from the computer-aided design file without the aid of any part-specific tooling. Throughout the preceding three decades, the number of three-dimensional printing techniques has significantly increased, changing the concept of direct printing of parts for a variety of applications. Three-dimensional printing technology provides key benefits for biomedical devices as well as tissue engineering because of its capability of manufacturing lower-volume or one-of-a-kind parts on a demand basis depending on patient-specific requirements, at zero additional cost for several designs which may vary from patient to patient, and while additionally offering flexibility in the initially used materials. Hence, there are still several issues that need to be resolved before 3-D printed biomaterials may be used widely. These issues include regulatory issues, the need for a sterile environment for part production, and achieving the target material properties with the appropriate architecture (Bandyopadhyay et al., 2015).

The printing technologies opened greater windows of creativity and innovation to biomaterials engineers by offering them the opportunity for fabricating complex shapes at a reasonable cost, time and weight. However, there was always a problem with the adjustment of function in printing technologies in view of a multiplicity of the materials as well as apparatus parameters. By transforming a digital subject into

DOI: 10.1201/9781003428862-6

a printed object (scaffolds, implants or diagnostic as well as drug delivery systems/devices), 3-dimensional printing, also termed as additive manufacturing, revolutionised the field of biomaterials engineering. In response to the requirement for the responsiveness of the printed platforms to a stimulus (such as temperature, light, pH, humidity, voltage, etc.) in a programmable manner, the concept of 4-D printing (majorly known as shape-morphing fabrication) was conceptualised and also put into practice. Later, 5-dimensional printing, which allowed for the printing of desired objects from five axes as opposed to 3-D printers' one-point printing upward, marked the next major advancement in printing technology. In comparison, 5-dimensional printers utilise 20–30% fewer materials, making it possible to create curved surfaces. However, a bioink with the right qualities for biomedical applications is required for all bioprinters. So, we briefly touched on the affordability, biomimicry of the scaffold, biodegradation of the scaffold, and cell viability (Shokrani et al., 2022).

The field of biofabrication has seen a revolution since the introduction of additive manufacturing, also termed 3-dimensional printing, which has also sparked numerous important developments in tissue engineering and regenerative medicine. After Charles Hull first made known to the world about the potential of stereolithography, a variety of 3-dimensional printing techniques were created in a short period of time. The field's obstacle, however, was materials development, which was not received with the same excitement. The materials toolbox for 3-dimensional printing applications has only recently been consciously expanded to reach the full potential of 3-dimensional printing technology. With a highlight on bioprinting applications, in this chapter we examine the production of the biomaterials appropriate for a light-based 3-dimensional printing modalities. We go over the chemical processes that control photopolymerization and emphasise the use of organic, inorganic and composite biomaterials in the creation of 3-D printed hydrogels (Yu et al., 2020). The characterisation and processing of advanced materials before and after the 3-D printing process have been well noted and show the effective influence of parametric conditions (Sharma et al., 2022; Sharma & Jain, 2020; Sharma et al., 2018; Sharma et al., 2019a; Sharma et al., 2019b; Sharma et al., 2019c; Sharma et al., 2021a; Sharma et al., 2021b; Sharma et al., 2020; Sharma et al., 2022; Kalman et al., 2017; Lumay et al., 2019; Tripathi & Mallick, 2017; Neveu et al., 2020a; Neveu et al., 2020b; Ratsimba et al., 2021; Tripathi et al., 2020; Tripathi et al., 2019; Tripathi et al., 2016; Tripathi & Mallick, 2014).

Because of its advantages, including custom fabrication, lesser costs of production, unparalleled capabilities for complicated geometry and quicker fabrication time, the 3-dimensional printing technology attracted interest in the field of business and academia during the past few decades. A cutting-edge technology which can permit the production of metallic implants for biomedical applications is 3-dimensional printing of metals having programmable shapes (Ni et al., 2019).

Innovations in the invention of new delivery systems for pharmaceutical products were continuously taking place in an effort to increment the therapeutic efficacy of the medications. Customised, personalised medications have been produced with the help of 3-dimensional printing technology to maximise therapeutic effects for the patients. In order to form a paradigm change in the areas of the healthcare sector,

3-dimensional printing has also been employed to create drug delivery systems and biological devices. We present an update on recent developments, technological gaps, and legal issues related to 3-dimensional printing technology in this assessment (Beg et al., 2020).

The development of high-value biomaterials from marine resources has enormous promise. Biomaterials of marine origin have become more prevalent during the past ten years. This industry is developing quickly. Coralline bone graft and polysaccharide-based biomaterials are just two examples of marine biomaterials that have gone through various stages of discovery and development. Chitin as well as chitosan, the collagen generated from marine species, and composites made of various marine organisms are examples of the latter. These materials are readily available and have strong bioactive properties, great biocompatibility, and biodegradability. Medical applications, drug delivery agents, antibacterial agents, anticoagulants, rehabilitation of diseases like diabetes, cardiovascular diseases, and bone disorders, the edible as well as aesthetic uses, are important applications of marine biomaterials (Wan et al., 2021).

Nature has evolved species over millions of years that perform astoundingly well because of their distinctive materials and structures, giving us invaluable inspiration for the designing of the next-generation biomedical equipment. The manufacture of multiscale, multi-functional and multi-material, three-dimensional (3-D) biomimetic materials and structures having higher precision as well as significant flexibility is made possible by the exciting new technology of 3-dimensional printing. With the burgeoning developments of 3-dimensional printing technology, the manufacturing challenges of biomedical devices having cutting-edge biomimetic materials as well as structures for numerous purposes were overcome. Medicine, implants, lab-on-a-chip, microvascular networks and artificial organs along with tissues were among the various biomedical uses that were discussed. Further research was done on the technical difficulties and constraints of biomimetic additive manufacturing in biomedical applications, and it was also done to highlight any potential solutions and fascinating next-generation technological advancements for biomimetic 3-dimensional printing of biomedical devices (Zhu et al., 2021).

In order to create orthoses, scaffolds, and prosthetic devices for regenerative medicine, tissue engineering, and rehabilitation for patients with neurological diseases (such as traumatic brain injuries, amyotrophic lateral sclerosis, and spinal cord injuries), three-dimensional (3-dimensional) printing is becoming a booming technology. This is because 3-dimensional printing has the ability to offer patient-specific designs, very complicated structural designs and quick, on-demand manufacture at a cheap cost. The absence of polymers, hydrogels, bioinks, and biomaterials, that are suitable for three-dimensional (3-D) biocompatible printing, and being more effective from a biomechanical point of view is one of the main obstacles preventing the widespread adoption of 3-dimensional printing for biomedical manufacturing. The processing of such materials into self-supporting devices carrying adjustable biomechanics, ideal architectures, degradation, and bioactivity is still a challenge for the field. Here, all new developments in 3-dimensional printing for biomedicine will be discussed. Polymers, hydrogels, and bioinks will be evaluated for printability, cost, simplicity of processing, and other qualities including mechanics, biocompatibility, and degradation rate (Pugliese et al., 2021).

In order to achieve positive clinical outcomes, reconstructive surgery seeks to correct tissue abnormalities by replacement with equivalent autologous tissue. For the restoration of the anatomical shape as well as partial function, however, synthetic materials are frequently needed because the lesion is too wide or there is insufficient tissue. The use of three-dimensional (3-D) printing enables the production of implants with intricate internal structures that more closely resemble the necessary therapeutic requirements. As biomedical materials, synthetic polymers have some advantages over natural polymers because they can more accurately imitate the mechanical and chemical characteristics of biological tissue. Because synthetic polymer materials are flexible in their chemical and mechanical properties as well as their physical form, 3-dimensional printing can produce 3-dimensional objects from materials like poly (lactic acid) as well as acrylonitrile butadiene styrene with ease. Due to their excellent mechanical qualities, biocompatibility as well as hemocompatibility, polyurethanes (PUs) are frequently employed as short- and long-term implantable medical devices. An overview of the development of 3-dimensional printable PU-based materials for biomedical applications is given in this chapter. To clarify how polyurethanes (Pus) may be manufactured into medical devices utilising additive manufacturing techniques, an overview of the chemical makeup along with the synthesis of PUs is given. The use of PUs in various 3-dimensional printing techniques, such as fused filament fabrication, bio-plotting, and stereolithography is currently being investigated to create complicated implants with accurate shapes and patterns with fine resolution. Three-dimensional printed PU scaffolds have demonstrated good in vivo tissue integration and cell survival. To encourage further research, the significant limitations of polyurethanes (PUs) printing are identified. In order to create implants that are designed to fulfil particular mechanical, anatomical, and biological criteria for biomedical applications, PUs provide a synthetic, biocompatible polymeric material which can be three-dimensional printed (Griffin et al., 2020).

Over the past ten years, the use of additive manufacturing or 3-dimensional printing, has increased significantly in the field of tissue engineering due to the invention of numerous new printing materials. Materials are employed in extrusion-based printing for a variety of purposes, from cell-free printing to cell-filled bioinks that resemble genuine tissues. Multi-material extrusion-based printing is currently being developed to form scaffolds which simulate tissue interfaces in addition to the single tissue applications. Despite these developments, some material restrictions limit the widespread use of the current extrusion-based 3-dimensional printers. And this progress report gives an overview of the widely used printing strategy and offers suggestions on how it may be enhanced. In order to facilitate cross-platform usage and reproducibility, it is intended that the anticipated report would serve as a guide for the inclusion of a more thorough material characterization prior to printing (Placone et al., 2018).

Biomaterials which can interface with biological systems are applicable in constructing functional tissues outside of the body for the replacement of organs, delivering medications safely and effectively, preventing, detecting, and curing diseases. With a primary focus on qualities that allowed for function restoration and acute pathology mitigation, the field has advanced beyond choosing materials that were initially intended for other applications. In order to integrate with biological complexity and carry out customized, high-level functions in the body, biomaterials are currently logically constructed with regulated structure and also dynamic functionality. The change has been from being permissive to encouraging biomaterials that are now bioactive rather than bioinert. With a focus on improvements in nano- to that of macroscale control, the static to the dynamic functionality, and the bio-complex materials, this perspective examines current advancements in the areas of polymeric as well as soft biomaterials (Tibbitt et al., 2015).

The ideal method for computer-aided production of intricate geometries for cutting-edge technology is 3-D printing. Designing criteria are painstakingly optimised in 3-dimensional printing processes due to the exact dimensions and shapes of printed products. However, producing 3-dimensional printable polymers has always been hampered by their poor mechanical properties, which called for the addition of reinforcing agents. Usually, neat polymers are unable to meet the demands of the ultimate application. Nanomaterials are used to improve the characteristics of polymers. Materials should be carefully chosen based on the intended use in order to produce a printed material with the proper qualities. Nanocomposites that have been 3-D printed are considered for use in the biomedical and electronic industries. Numerous scientific reports have focused on designing 3-dimensional printed nanocomposites as a biomaterial, primarily for medical applications. It was acknowledged that creating suitable nanocomposites for 3-dimensional printing requires acquiring extensive knowledge of materials' structures and rheological characteristics. However, a few reports focused on the classification of reports on 3-dimensional printed biomaterials for medical applications. This chapter reviews 3-dimensional printed nanocomposites in light of their mechanical and rheological characteristics for use in biomaterials (Zarrintaj et al., 2021).

A major structural protein that is present in the connective tissues, like tendons, bone, skin, and cartilage in animals, is collagen. Natural collagen has a multi-hierarchical fibrous architecture as a result of its complex manufacture in vivo, which involves a lot of intracellular as well as extracellular stages. Collagen's bioactivity primarily depends on its tertiary structure or higher. Collagen biomaterials have drawn a lot of attention in recent years for use in biological applications because of their great qualities, including lesser immunogenicity, biodegradability, biocompatibility, hydrophilicity, and ease of processing. However, collagen also has poor physical and chemical characteristics (thermostability, mechanical strength, enzyme resistance, etc.). As a result, collagen must be modified during the preparation process. The crosslinking techniques and most current developments of collagen-based materials in biomedical applications, such as skin substitute, tendon, bone, neural, cartilage also the delivery system, will be discussed in this study (Zhu et al., 2016).

Because of its exceptional qualities, including quick recovery, self-healing, biocompatibility and higher mechanical properties mixed with multi-stimulus responsiveness, metal ion cross-linked hydrogels are currently attracting attention. We have outlined the most recent developments in the creation of metal ion cross-linked hydrogels for the engineering of tissue and biomedical applications in this chapter. Regarding the cross-linking mechanisms, physio-chemical, compositions, and biological properties of various cross-linked hydrogels, a lot of metal ions, and their role in their synthesis are discussed. Ferric ($Fe3+$) ion cross-linked hydrogels along with their various combinations have received special attention due to the numerous studies that have recently been published on them and their outstanding features. With particular examples, the use of these metal ion-based hydrogels in biomedical applications is covered, such as sensing, medication delivery, engineering of tissue, healing of wounds, as tissue adhesives, and also sealants. It is significant to note that a separate section of the chapter explains how these metal ion cross-linked hydrogels can be used as inks in 3-dimensional printing. The potential toxicity of the several metal ions and their consequences have also been carefully examined. The hydrogels' potential future uses and wide-ranging applications are underlined. For use in biomedical engineering, particularly tissue engineering, hydrogels are the 3-dimensional networks made of cross-linked polymer chains that can absorb water while also accommodating various medications and bio-factors. Patients around the world have tissue damage or organ failure which requires an abundance of tissues or organs to be replaced. However, many people continue to wait for the right transplant due to the lack of donors. The influence of hydrogels and their constituent parts shouldn't have an impact on the cells or other components in tissue engineering or other biomedical applications. The extracellular matrix (ECM) as well as hydrogels is comparable in that both offer a 3-dimensional environment for the cells to attach, grow, proliferate, and differentiate (Janarthanan et al., 2021).

For the past 40 years, additive manufacturing (AM) has evolved into an efficient on-demand technique for creating geometrically complex items. Due to its ability to design and print practically any object shape using a larger variety of materials, including polymers, metals, bioinks, and ceramics, this approach has been embraced for biomedical applications in both research and clinical settings. Additive manufacturing will continue to play a major role in healthcare as a result of current developments in medical device manufacturing, therapeutic delivery, tissue engineering and regeneration, and also in operative management planning (Ahangar et al., 2019).

A range of engineering, scientific, and medical fields have benefited greatly from 3-dimensional printing, which is an area of study that is rapidly increasing. The scientific improvement of 3-dimensional printing technologies has made it feasible to create complicated geometries, but there is still an incrementing value for new 3-dimensional printing materials along with the methods to solve problems with building speed and accuracy, surface smoothness, stability, and functionality. This chapter introduces and examines the most recent developments in cutting-edge materials as well as 3-dimensional printing technologies to address the requirements of a conventional 3-dimensional printing methodology, particularly in biomedical applications such as cell growth feasibility, printing speed, and complicated form achievement. Comparison research of various materials and methods in connection with the 3-dimensional printing parameters will be provided to select an acceptable application-based 3-D-printing technique (Tetsuka et al., 2020).

Three-dimensional (3-D) bioprinting, also termed as an additive manufacturing-based method of fabricating biomaterials, is a cutting-edge technique along with promising tactics in the medical as well as pharmaceutical industries. This technique is a pioneer in the development of artificial multicellular tissues and organs due to its capacity to produce regenerative tissues and organs. Currently, 3-dimensional bioprinting is being used by researchers applying a broader range of biomaterials and various techniques. In this chapter, we further illustrate the most prevalent and cutting-edge biomaterials used in 3-D bioprinting technology while introducing the associated approaches which were frequently considered by the researchers. Additionally, by showcasing the most significant works, an effort has been made to convey the major pertinent applications of the 3-dimensional bioprinting process, like the regeneration of tissue, cancer research, etc. The major goal of this chapter is to focus on the drawbacks and benefits of existing biomaterials as well as 3-dimensional bioprinting technologies in order to make comparisons between them (Vanaei et al., 2021).

Because of recent advancements in biology, material sciences, and particularly a newly developed additive manufacturing process called three-dimensional (3-D) printing, the engineering of tissue as well as regenerative medicine entered a newer stage of development. A cutting-edge biofabrication method known as 3-D printing may create patient-specific scaffolds with extremely complicated shapes that host cells as well as bioactive substances to speed up tissue regeneration. Because of its wider availability, structural characteristics, and advantageous biological qualities, chitosan hydrogels have been employed extensively in a lot of biomedical applications; however, 3-dimensional printing chitosan-based hydrogels is still in its infancy. In order to investigate the possible use of chitosan as a 3-dimensional printing ink or as a coating on the other 3-dimensional printed scaffolds, 3-dimensional printing technologies present a new path. Next-generation biomedical implants have a lot of potential thanks to the applicability of hydrogels based on chitosan as well as 3-D printing (Rajabi et al., 2021).

Over the previous 10 years, the developments in additive manufacturing have sparked the formation of cutting-edge tactics across a number of healthcare industries. Because of inadequate healing, increase in patients, and the financial strain on the healthcare systems, the treatment as well as healing of chronic wounds may continue to present significant clinical challenges. The formation of cell-laden bioinks, the application of novel synthetic or natural biomaterials, and the utilisation of pharmacological substances paved the way for the efficient management of wounds as well as therapy, including the engineering of skin substitutes and regeneration of skin. The key advantages of printing technologies include the capability of combining an infinite number of bioactive chemicals as well as cells with polymers, the ability to create intricate scaffold designing, quicker healing times, and customised wound dressings. Figure 6.1 illustrate the theranostics applications of 3D cell printing.

The concept of additive manufacturing procedure utilised for wound healing is highlighted in this chapter, along with technological benefits and processing restrictions. The applicability of the materials having antibacterial, antioxidant or the integration of pharmacological compounds as well as peptides for the treatment of chronic wounds is also covered, along with current trends and clinical advancements

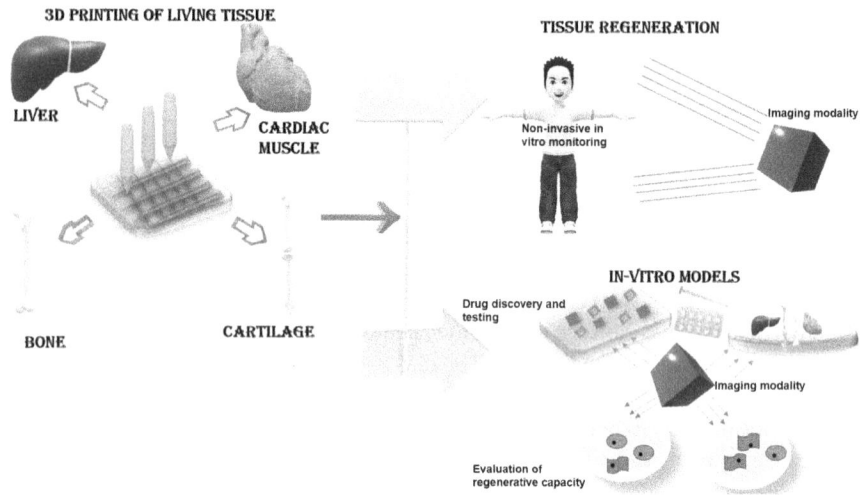

FIGURE 6.1   Schematic illustration for theranostics applications of 3-D cell printing.

in this area. Finally, we offer 3-D-printing-based methods for tissue engineering of skin as well as regeneration.

6.2   TYPES OF BIOMATERIAL INKS USED IN 3-D PRINTING

Biomaterial inks are able to meet different traditional biomaterial needs (biodegradability, biocompatibility, etc.) in addition to the 3-dimensional printing requirements. Consequently, a lot of biomaterial printing work has been performed towards the development techniques of 3-dimensionally printed biomaterials as compared to the previously developed materials, that may hold polyesters, hydrogels, and polymer-ceramic composites.

6.2.1   HYDROGELS

Hydrogels were printed with the help of several processes: most importantly the layer-by-layer cross-linking, thickened solutions (mainly higher polymer fractions), as well as lesser often smallest additions of the cross-linkers. Some examples are the agar or agarose, the alginate, the PEG derivatives, and the poloxamers (Pluronic®, Lutrol®). The hydrogel inks were printed having better compatibility of cells, yet they were un-modified and were termed to as 'blank-slate' materials, which may lack the adhesion of cells as well as the cell degradation (Jakus et al., 2016).

6.2.2   CERAMIC-BASED INKS

Currently, one of the key beneficial strategies in the area of bone tissue engineering has been focusing on the production of biomimetic scaffolds. The ceramic-based scaffolds along with the favourable osteogenic capability as well as mechanical properties are termed to be promising candidates for the repairing of bones. The three-dimensional (3-D) printing is found to be an additive manufacturing method that can allow the fabrication of patient-specific scaffolds having higher structural complexities and design flexibilities, as well as gaining increasing attention (Du et al., 2018).

6.2.3 POLYMER-BASED INKS

Bolus is a type of auxiliary device applied in radiotherapy for the treatment of superficial lesions like skin cancer. It is commonly applicable to incremented skin dose as well as to overcome the skin-sparing effect. Despite the availability of several commercial boluses, there is at present no bolus that can form full contact with the irregular surface of the patient's skin, and incomplete contact would result in air gaps. The resulting air gaps had decreased the surface radiation dose, leading to a discrepancy between a planned dose and a delivered dose. To reduce this limitation, the customized bolus processed by 3-dimensional (3-D) printing carries tremendous potential for developing radiotherapy more efficiently than ever before. An ideal material for the customizing bolus must not only have the feature of 3-dimensional printability for customization but also possess the radiotherapy adjuvant performance as well as other multiple compound properties, such as antibacterial activity, biocompatibility, tissue equivalence, and anti-phlogosis (Lu et al., 2021).

6.2.4 COMPOSITE INKS

Most of the time, bone injuries are repaired without any consequences, but there are an increasing number of situations where significant therapeutic intervention is necessary for bone healing. Availability of treatment options shows several disadvantages, such as donor site morbidity as well as limited availability for the process like autografting. Bone graft substitutes carrying growth factors would be a viable alternative, hence they have been associated with dose-related safety concerns as well as also lack the control over spatial architecture to match the bone defect sites anatomically.

Three-dimensional printing provides a solution to make patient-specific bone graft substitutes which are customized to the patient's bone defect having temporal control over incorporated therapeutics for maximizing their efficacy. As inspired by the natural constitution of the bone tissue, composites made of the inorganic phases, like as calcium phosphate, nano-silica particles, and bioactive glasses, combined with the biopolymer matrices have been investigated as the building blocks for the biofabrication of the bone constructs. Besides capturing the elements of the bone physiological structure, these organic/inorganic composites must be designed for a specific cohesivity, the rheological as well as mechanical properties, and hence both organic as well as inorganic constituents may contribute to composite bioactivity (Van der Heide et al., 2022).

6.3 APPLICATION OF 3-DIMENSIONAL PRINTING IN BIOMEDICALS

6.3.1 DENTAL IMPLANTS

The formation of 3-dimensional printing implantable medical devices is merely related to the production of the medical level and material science, mainly the production of the biomaterials. The biological materials are substances which must be applied to diagnose, repair, treat, or replace the organs as well as tissues in the body or increment their functions. And they must be artificial, natural, or a combination of both. The biomaterials applicable in the 3-dimensional printing of implantable medical devices may be separated into biomedical polymer materials, biomedical ceramic materials, biomedical metal materials, biomedical composite materials, and derived materials according to their properties (Wang & Yang 2021).

6.3.2 BIO-PRINTING

Utilisation of stem cell technology, scaffolding design, and nanotechnology have opened novel opportunities in the regeneration of tissue. The application of accurate engineering designing in the production of scaffolds, such as 3-dimensional printers were considered widely. Three-dimensional printers, mainly higher precision bio-printers, may create a novel way in the designing of 3-dimensional tissue engineering scaffolds.

6.3.3 DRUG DELIVERY

Production of pharmaceutical products is continuously innovating the designing of newer delivery systems to improve the therapeutic efficacies of the drugs. Three-dimensional printing technologies have been applied in designing the customized personalized medication for providing higher therapeutic advantages for patients. In addition, 3-dimensional printing has also been applicable in the manufacturing of the drug delivery systems as well as biomedical devices in establishing a paradigm shift in the healthcare industries (Prasad et al., 2016).

6.4 CONCLUSION

Biomaterials have experienced significant advancements in the current years. They are also identified as being one of the emerging technologies that have revolutionized the areas called 3-dimensional printing. Three-dimensional printing, also termed additive manufacturing, provides for the precise fabrication of complex structures layer by layer, providing tremendous potential for the applications in biomedical. And one of the key advantages of 3-dimensional printing in the context of biomaterials is their availability for creating patient-specific implants as well as devices.

The traditional method often relies on standardized shapes and sizes that may not fit perfectly in patients. With 3-dimensional printing, it is possible to generate personalized implants that can match the anatomical specifications of a patient, resulting in the improvement of compatibility and functionality. Furthermore, 3-dimensional printing may offer the fabrication of intricate structures carrying controlled micro-architectures. This capability is mainly valuable in tissue engineering as well as in regenerative medicine, where the main goal is to develop biomimetic scaffolds which may help in the growth of the cell and regeneration of the tissues.

By precisely controlling the spatial arrangements of the materials, 3-dimensional printing may allow for the formation of the scaffolds having a specific mechanical property, porosity as well as bioactive cues to help the cellular behaviour as well as tissue integration. In addition to the formation of the automatically precise implants as well as the tissue scaffolds, 3-Dimensional printing also facilitates the formation of the drug delivery systems having tailored characteristics. This technology has been applicable to fabricate complex drug release profiles, such as multi-layered systems or devices having compartments for the subsequent release. By precisely controlling the special distribution as well as the release of the kinetics of the drugs, the 3-dimensional printing drug delivery system may increment the therapeutic outcomes as well as minimize the side effects.

Another sector where the 3-dimensional printing is making significant strides is in the formation of bioactive as well as biodegradable materials. The researchers were exploring the application of bioinks, which are printable biomaterials having growth factors, living cells, and other bioactive components. These bioinks must be applicable in printing functional organs and tissues, providing potential solutions for personalized medicine and for organ transplantation. Despite the tremendous promises of 3-dimensional printing in the areas of biomaterials, there are still various challenges that need to be focused. These hold the development of mechanically robust and biocompatible materials which are suitable for the 3-dimensional printing, to ensure the proper cell viability as well as integration within the printed structures and establishing the standardized protocols along with the regulatory framework for the clinical transplantation of the 3-dimensional-printed biomedical devices. In conclusion, 3-dimensional printing carries greater potential for the next generation of biomedical and biomaterial applications.

With further research and advancements, it is anticipated that technologies may drive innovations in the engineering of tissues, personalized medicines, drug delivery and regenerative medicines, ultimately leading to an advanced patient care as well as outcomes.

REFERENCES

Ahangar, P., Cooke, M. E., Weber, M. H., & Rosenzweig, D. H. (2019). Current biomedical applications of 3D printing and additive manufacturing. *Applied sciences*, 9(8), 1713.

Bandyopadhyay, A., Bose, S., & Das, S. (2015). 3D printing of biomaterials. *MRS bulletin*, 40(2), 108–115.

Beg, S., Almalki, W. H., Malik, A., Farhan, M., Aatif, M., Rahman, Z., Alruwaili, N. K., Alrobaian, M., Tarique, M., & Rahman, M. (2020). 3D printing for drug delivery and biomedical applications. *Drug Discovery Today*, 25(9), 1668–1681.

Du, X., Fu, S., & Zhu, Y. (2018). 3D printing of ceramic-based scaffolds for bone tissue engineering: An overview. *Journal of Materials Chemistry B*, *6*(27), 4397–4412.

Griffin, M., Castro, N., Bas, O., Saifzadeh, S., Butler, P., & Hutmacher, D. W. (2020). The current versatility of polyurethane three-dimensional printing for biomedical applications. *Tissue Engineering Part B: Reviews*, *26*(3), 272–283.

Jakus, A. E., Rutz, A. L., & Shah, R. N. (2016). Advancing the field of 3D biomaterial printing. *Biomedical Materials*, *11*(1), 014102.

Janarthanan, G., & Noh, I. (2021). Recent trends in metal ion based hydrogel biomaterials for tissue engineering and other biomedical applications. *Journal of Materials Science & Technology*, *63*, 35–53.

Kalman, H., Tripathi, N. M., Gabrieli, O. G., & Portnikov, D. (2017). Phase diagrams for pneumatic and hydraulic conveying. In *18th International Conferences on Transport and Sedimentation of Solid Particles, T and S 2017* (pp. 145–152). Wydawnictwo Uniwersytetu Przyrodniczego we Wroclawiu.

Lu, Y., Song, J., Yao, X., An, M., Shi, Q., & Huang, X. (2021). 3D printing polymer-based bolus used for radiotherapy. *International Journal of Bioprinting*, *7*(4).

Lumay, G., Tripathi, N. M., & Francqui, F. (2019). How to gain a full understanding of powder flow properties, and the benefits of doing so. *ONdrugDelivery*, *2019*, 42–47.

M Tripathi, N., & S Mallick, S. (2017). Pneumatic conveying of Fly Ash: Bend Models investigation. *Advanced Materials Proceedings*, *2*(8), 526–531.

Neveu, A., Lumay, G., Pillitteri, S., Monsuur, F., Pauly, T., Ribeyre, Q., Francqui, F., Vandewalle, N., & Tripathi, N. M. (2020b). Physical characterization of blends containing mesoporous particles with a focus on electrostatic properties. In *2020 AIChE Spring Meeting and 16th Global Congress on Process Safety*. AIChE.

Neveu, A., Tripathi, N. M., Rigo, O., Francqui, F., & Lumay, G. (2020a). Experimental investigation of spreadability of metal powders in recoating process. In *Proceedings - Euro PM2020 Congress and Exhibition (Proceedings - Euro PM2020 Congress and Exhibition)*. European Powder Metallurgy Association (EPMA).

Ni, J., Ling, H., Zhang, S., Wang, Z., Peng, Z., Benyshek, C., Zan, R., Miri, A. K., Li, Z., Zhang, X., Lee, J., Lee, K-J., Kim, H-J., Tebon, P., Hoffman, T., Dokmeci, M. R., Ashammakhi, N., Li, X., & Khademhosseini, A. (2019). Three-dimensional printing of metals for biomedical applications. *Materials Today Bio*, *3*, 100024.

Placone, J. K., & Engler, A. J. (2018). Recent advances in extrusion-based 3D printing for biomedical applications. *Advanced Healthcare Materials*, *7*(8), 1701161.

Prasad, L. K., & Smyth, H. (2016). 3D Printing technologies for drug delivery: A review. *Drug Development and Industrial Pharmacy*, *42*(7), 1019–1031.

Pugliese, R., Beltrami, B., Regondi, S., & Lunetta, C. (2021). Polymeric biomaterials for 3D printing in medicine: An overview. *Annals of 3D Printed Medicine*, *2*, 100011.

Rajabi, M., McConnell, M., Cabral, J., & Ali, M. A. (2021). Chitosan hydrogels in 3D printing for biomedical applications. *Carbohydrate Polymers*, *260*, 117768.

Ratsimba, A., Zerrouki, A., Tessier-Doyen, N., Nait-Ali, B., André, D., Duport, P., Neveu, A., Tripathi, N., Francqui, F., & Delaizir, G. (2021). Densification behaviour and three-dimensional printing of Y2O3 ceramic powder by selective laser sintering. *Ceramics International*, *47*(6), 7465–7474. https://doi.org/10.1016/j.ceramint.2020.11.087

Sharma, A., Babbar, A., Tian, Y., Pathri, B. P., Gupta, M., & Singh, R. (2022). Machining of ceramic materials: A state of the art review. *International Journal on Interactive Design and Manufacturing (IJIDeM)*, 1–21. https://doi.org/10.1007/s12008-022-01016-7

Sharma, A., & Jain, V. (2020). Experimental investigation of cutting temperature during drilling of float glass specimen. In *IOP Conference Series: Materials Science and Engineering* (Vol. 715, No. 1, p. 012050). IOP Publishing.

Sharma, A., Jain, V., & Gupta, D. (2018). Characterization of chipping and tool wear during drilling of float glass using rotary ultrasonic machining. *Measurement, 128*, 254–263.

Sharma, A., Jain, V., & Gupta, D. (2019a). Tool wear analysis while creating blind holes on float glass using conventional drilling: A multi-shaped tools study. In *Advances in Manufacturing Processes: Select Proceedings of ICEMMM 2018* (pp. 175–183). Springer Singapore.

Sharma, A., Jain, V., & Gupta, D. (2019b). Comparative analysis of chipping mechanics of float glass during rotary ultrasonic drilling and conventional drilling: For multi-shaped tools. *Machining Science and Technology, 23*(4), 547–568.

Sharma, A., Jain, V., & Gupta, D. (2019c). Multi-shaped tool wear study during rotary ultrasonic drilling and conventional drilling for amorphous solid. *Proceedings of the Institution of Mechanical Engineers, Part E: Journal of Process Mechanical Engineering, 233*(3), 551–560.

Sharma, A., Jain, V., & Gupta, D. (2021a). Effect of pre and post tempering on hole quality of float glass specimen: For rotary ultrasonic and conventional drilling. *Silicon, 13*, 2029–2039.

Sharma, A., Jain, V., & Gupta, D. (2021b). Mathematical approach on chipping volume estimation generated during rotary ultrasonic drilling for float glass. *Proceedings of the National Academy of Sciences, India Section A: Physical Sciences, 92*, 285–291.

Sharma, A., Jain, V., Gupta, D., & Babbar, A. (2020). A review study on miniaturization. In *Advanced Manufacturing and Processing Technology* (First edition, pp. 111–131). CRC Press.

Shokrani, H., Shokrani, A., & Saeb, M. R. (2022). Methods for biomaterials printing: A short review and perspective. *Methods*.

Tetsuka, H., & Shin, S. R. (2020). Materials and technical innovations in 3D printing in biomedical applications. *Journal of Materials Chemistry B, 8*(15), 2930–2950.

Tibbitt, M. W., Rodell, C. B., Burdick, J. A., & Anseth, K. S. (2015). Progress in material design for biomedical applications. *Proceedings of the National Academy of Sciences, 112*(47), 14444–14451.

Tripathi, N. M., Francqui, F., & Lumay, G. (2020). Influence of relative air humidity on the flow property of fine powders. In *Third International Conference on Powder, Granule and Bulk Solids: Innovations and Applications PGBSIA 2020 February 26–28, 2020* (p. 63).

Tripathi, N. M., Francqui, F., Pirenne, T., & Lumay, G. (2019). Measuring food powders electrical properties as a result of anti-static content. In *2019 AIChE Annual Meeting*. American Institute of Chemical Engineers.

Tripathi, N. M., Levy, A., & Kalman, H. (2016). Initial acceleration pressure drop in dilute phase pneumatic conveying system. In *Powder, Granule and Bulk Solids: Innovations and Applications Conference*.

Tripathi, N. M., & Mallick, S. S. (2014). *An Investigation into Pressure Drop Across Bends for Fluidised Densephase Pneumatic Conveying Systems* (Doctoral dissertation).

van der Heide, D., Cidonio, G., Stoddart, M. J., & D'Este, M. (2022). 3D printing of inorganic-biopolymer composites for bone regeneration. *Biofabrication, 14*(4), 042003.

Vanaei, S., Parizi, M. S., Salemizadehparizi, F., & Vanaei, H. R. (2021). An overview on materials and techniques in 3D bioprinting toward biomedical application. *Engineered Regeneration, 2*, 1–18.

Wan, M. C., Qin, W., Lei, C., Li, Q. H., Meng, M., Fang, M., Song, W., Chen, J-H., Tay, F., & Niu, L. N. (2021). Biomaterials from the sea: Future building blocks for biomedical applications. *Bioactive Materials, 6*(12), 4255–4285.

Wang, Q., Han, G., Yan, S., & Zhang, Q. (2019). 3D printing of silk fibroin for biomedical applications. *Materials, 12*(3), 504.

Wang, Z., & Yang, Y. (2021). Application of 3D printing in implantable medical devices. *BioMed Research International, 2021*, 1–13.

Yu, C., Schimelman, J., Wang, P., Miller, K. L., Ma, X., You, S., Guan, J., Bingjie, S., Zhu, W., & Chen, S. (2020). Photopolymerizable biomaterials and light-based 3D printing strategies for biomedical applications. *Chemical Reviews, 120*(19), 10695–10743.

Zarrintaj, P., Vahabi, H., Gutiérrez, T. J., Mehrpouya, M., Ganjali, M. R., & Saeb, M. R. (2021). Nanocomposite biomaterials made by 3D printing: Achievements and challenges. In *Handbook of Polymer Nanocomposites for Industrial Applications* (pp. 675–685). Elsevier.

Zhu, W., Ma, X., Gou, M., Mei, D., Zhang, K., & Chen, S. (2016). 3D printing of functional biomaterials for tissue engineering. *Current Opinion in Biotechnology, 40*, 103–112.

Zhu, Y., Joralmon, D., Shan, W., Chen, Y., Rong, J., Zhao, H., Xiao, S., & Li, X. (2021). 3D printing biomimetic materials and structures for biomedical applications. *Bio-Design and Manufacturing, 4*, 405–428.

7 Biomedical Innovations with 3D Printing and Biomaterials
Current and Future Applications

Adhishree Yadav, Vikas Raghuvanshi,
Pramod Yadav, Vivek Mani Tripathi,
Samim Ali, and Deepak Singh Chauhan

7.1 INTRODUCTION

Over time, the manufacturing of medical equipment and accessories has evolved in response to changing environments and advancements in technology (Figure 7.1). Biomaterials and 3D printing have emerged as significant innovations in this field. The first commercial bioprinting technology, known as the "3D-Bioplotter," was introduced by Landers et al. (2002) in the early 21st century; however, the origins of 3D printing were in the late 19th century (1984) when Charles Hull developed and presented the first 3D printing technique, known as stereolithography (SLA). In 1988, Klebe et al. (1988) demonstrated the first biomedical use of a 3D printer based on cytoscribing technology. In 1996, Gabor Forgacs showed that tissue cohesion depends on molecular adhesion and can be measured by tissue surface tension (Foty et al., 1996). In 1999, Odde et al. utilized laser-assisted bioprinting to deposit living cells, enabling the development of analogues with complex anatomical structures. In 2001, the direct printing of a scaffold in the shape of a bladder and seeding of human cells took place (Karzyński et al., 2018). In 2003, Thomas Boland et al. developed the first inkjet bioprinter by modifying a standard HP inkjet printer, and a year later, their team implemented cell-loaded bioprinting with a commercial SLA printer (Dhariwala et al., 2004; Wilson & Boland, 2003). In 2006, electrohydrodynamic jetting was used to deposit living cells (Jayasinghe et al., 2006). In 2009, Norotte et al. (2009) successfully developed the first scaffold-free bioprinted tissue, specifically vascular tissue. In 2012, Skardal attempted in situ bioprinting on mouse models and also

DOI: 10.1201/9781003428862-7

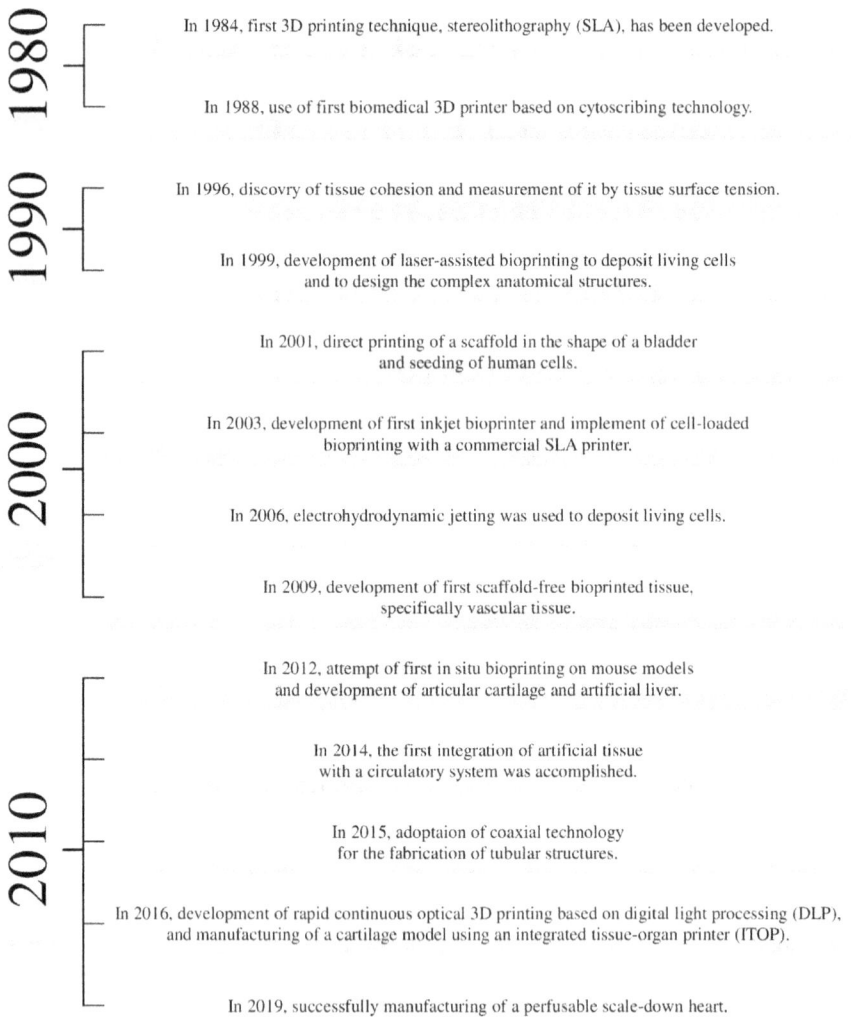

1980
In 1984, first 3D printing technique, stereolithography (SLA), has been developed.

In 1988, use of first biomedical 3D printer based on cytoscribing technology.

1990
In 1996, discovry of tissue cohesion and measurement of it by tissue surface tension.

In 1999, development of laser-assisted bioprinting to deposit living cells and to design the complex anatomical structures.

In 2001, direct printing of a scaffold in the shape of a bladder and seeding of human cells.

In 2003, development of first inkjet bioprinter and implement of cell-loaded bioprinting with a commercial SLA printer.

2000
In 2006, electrohydrodynamic jetting was used to deposit living cells.

In 2009, development of first scaffold-free bioprinted tissue, specifically vascular tissue.

In 2012, attempt of first in situ bioprinting on mouse models and development of articular cartilage and artificial liver.

In 2014, the first integration of artificial tissue with a circulatory system was accomplished.

In 2015, adoptaion of coaxial technology for the fabrication of tubular structures.

2010
In 2016, development of rapid continuous optical 3D printing based on digital light processing (DLP), and manufacturing of a cartilage model using an integrated tissue-organ printer (ITOP).

In 2019, successfully manufacturing of a perfusable scale-down heart.

FIGURE 7.1   Summary of key developments in 3D bioprinting.

achieved the development of articular cartilage and artificial liver using other bioprinting techniques (Dababneh & Ozbolat, 2014; Duan, 2017). In 2014, the first integration of artificial tissue with a circulatory system was accomplished (Gao et al., 2015). In 2015, Gao et al. adopted coaxial technology for the tubular structure's fabrication (Pyo et al., 2017). In 2016, based on DLP, which stands for digital light processing, rapid and continuous optical 3D printing was developed by Pyo et al. (2017), and Anthony Atala's research group manufactured a cartilage model using an ITOP which stands for integrated tissue-organ printer (Kang et al., 2016). These advancements have significantly contributed to the progress and potential of 3D printing in the field of biomedical applications. Bioprinting is a process that uses biomaterials with 3D printing to create or replicate parts that imitate natural tissues such as blood vessels and bones of our body. The hardware that is used for bioprinting is known as a bioprinter. There are mainly five types of 3D bioprinter used these days; fused deposition modelling (FDM), stereolithography (SLA), selective laser sintering (SLS), Binder Jetting, and Material Jetting (Ozbolat et al., 2017). 3D bioprinting has been rapidly growing and is widely used in various biomedical applications such as tissue regeneration, identifying and optimizing drugs, studying the pathogenesis, and inventing a new medical application (Kumar et al., 2021; Tetsuka & Shin, 2020). Although, despite its vast medical applications and development of biomaterials, some challenges remain such as biocompatibility, long-term stability, and the potential for immune responses or rejection (Shahrubudin et al., 2020). However, the future of biomaterials is promising and there are several potential areas of growth and development ahead (Kačarević et al., 2018; Sharma et al., 2022; Sharma & Jain, 2020; Sharma et al., 2018; Sharma et al., 2019a; Sharma et al., 2019b; Sharma et al., 2019c; Sharma et al., 2021a; Sharma et al., 2021b; Sharma et al., 2020; Sharma et al., 2022; Kalman et al., 2017; Lumay et al., 2019; Tripathi & S Mallick, 2017; Neveu et al., 2020a; Neveu et al., 2020b; Ratsimba et al., 2021; Tripathi et al., 2020; Tripathi et al., 2019; Tripathi et al., 2016; Tripathi & Mallick, 2014).

7.2   WHAT ARE THE KEY COMPONENTS OF 3D BIO-PRINTER?

A bioprinter is a device that can print living cells using biomaterials to create 3D structures that mimic natural tissues and organs. It is typically made of three main components (Ozbolat et al., 2017).

7.2.1   HARDWARE

The hardware is the mechanical part of the bioprinter that controls the movement and deposition of the bio-ink (Zhang et al., 2021). It can use different printing techniques such as syringe-based extrusion, etc. Although, the physical description of a bioprinter may vary depending on the specific model and technology used, here is a physical description of a typical bioprinter (*Bioprinters Standards – The Future of ASME Standards – ASME*, n.d.; Zhang et al., 2021) (Figure 7.2):

1) *Frame*: The bioprinter consists of a sturdy frame, usually made of metal or high-quality plastic, that provides stability and support for the entire system.
2) *Build Platform*: The build platform is a flat surface where the printing process takes place. It can be moved in three dimensions (X, Y, and Z axes) to precisely position the printing nozzle or syringe.
3) *Printing Nozzle/Syringe*: The bioprinter is equipped with a specialized printing nozzle or syringe that dispenses the bio-ink, a mixture of living

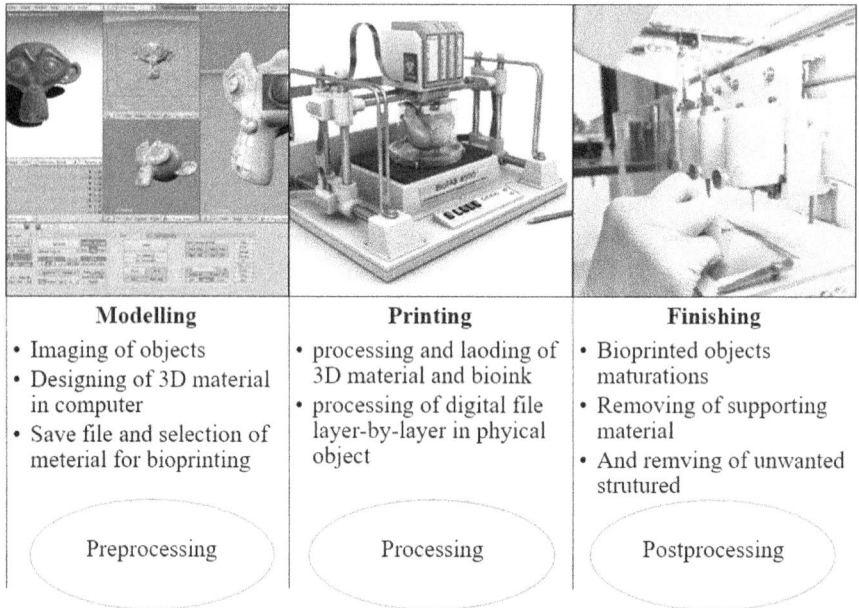

Modelling	Printing	Finishing
• Imaging of objects • Designing of 3D material in computer • Save file and selection of meterial for bioprinting	• processing and laoding of 3D material and bioink • processing of digital file layer-by-layer in phyical object	• Bioprinted objects maturations • Removing of supporting material • And remving of unwanted strutured
Preprocessing	Processing	Postprocessing

FIGURE 7.2   Presentation of 3D bioprinting steps.

cells and biomaterials, onto the build platform. The nozzle or syringe can be controlled to accurately deposit the bio-ink layer by layer.

4) *XYZ Motion System*: Bioprinters incorporate an XYZ motion system that enables precise movements of the build platform and printing nozzle. This system allows the bioprinter to create intricate structures with high resolution.

5) *Control Panel*: The bioprinter is equipped with a control panel or interface that allows the user to input printing parameters, such as layer thickness, speed, and temperature. It also displays real-time information about the printing process and provides options for customization.

6) *Enclosure*: To maintain a controlled environment, it often has an enclosure that helps regulate temperature, humidity, and sterility. This enclosure protects the printing process and ensures the viability of the printed structures.

7) *Filtration System*: It may feature a filtration system to remove impurities from the bio-ink and ensure a clean and sterile printing process. This helps to prevent contamination and maintain the integrity of the printed tissues or organs.

8) *Imaging System*: Some advanced bioprinters incorporate imaging systems, such as cameras or laser scanners, to monitor the printing process in real-time. These systems enable quality control and can provide feedback for adjustments during the printing process.

9) *Control Software*: Bioprinters are controlled by specialized software that translates the user's instructions into precise movements and commands for the printer components. The software also allows for design import, optimization, and customization of the printed structures.

7.2.2 BIO-INK

Bio-ink is the liquid mixture of cells, matrix, and nutrients that is printed by the bio-printer. It can be derived from various sources such as stem cells, adult cells, or synthetic cells, and different types of biomaterials (Table 7.1), such as natural or synthetic polymers, hydrogels, ceramics, or metals (Ng et al., 2017). The bioink needs to have certain properties such as viscosity, biocompatibility, and cell viability to be suitable for bioprinting. These properties affect the flow and deposition of bioink through the printing nozzle or needle, and the shape fidelity and stability of the printed structure (Donderwinkel et al., 2017). Some of the chemical properties of bioink include biocompatibility, bioactivity, degradation, and cross-linking. These properties affect the interaction and function of bio-ink with the biological environment and the living cells. Some of the other aspects of bio-ink include printability, biofunctionality, and biodegradability (Garcia-Cruz et al., 2021). Printability refers to the ability of bio-ink to be printed with high resolution, accuracy, and reproducibility. Biofunctionality refers to the ability of bio-ink to support the survival, growth, and differentiation of cells into functional tissues or organs. Biodegradability refers to the ability of bio-ink to degrade over time and be replaced by natural tissue (Naghieh & Chen, 2021). Recently a novel bioink was developed for the first time with the help of patient cells which allows small human-sized airways to be 3D-bioprinted (De Santis et al., 2021).

7.2.3 BIOMATERIALS

A biomaterial, also known as additive manufacturing, is referred to as a substance that has been created to be utilized in any therapeutic or diagnostic technique to control how one or more elements of biological systems interact when employed alone or as a component of a complicated device (*Biomaterials*, n.d.). It can be natural or synthetic polymers, ceramics, metals, or composites. Some biomaterials can also act as scaffolds. It is a structure that supports the bio-ink and provides mechanical strength, shape, and functionality to the bioprinted tissue or organ (Dey & Ozbolat, 2020). It also acts as a guide for the growth and differentiation of the cells in the bio-ink. It can be classified into different types based on their application, composition, or morphology. For example, it can be designed for soft tissue, hard tissue, or connective tissue engineering and also can be composed of organic, inorganic, polymeric, or natural materials (Dey & Ozbolat, 2020). Scaffolds have different shapes and sizes, such as hydrogel scaffolds, fibrous scaffolds, or porous scaffolds. It is important for bioprinting because they provide a suitable environment for the cell's survival, proliferation, and maturation into functional tissues or organs and maintains the shape and integrity of the bioprinted structure during and after the printing process (Liu et al., 2022).

TABLE 7.1

A Comparative Table of Different Types of Biomaterials Used in 3D Bioprinting

Material	Natural Polymers	Synthetic Polymers	Ceramics	Metals	Composites
Definition	Derived from natural sources, such as plants or animals, proteins, polysaccharides, nucleic acids, alginate.	Artificially synthesized from organic/inorganic monomers, such as polyethylene, polyvinyl alcohol, polylactic acid.	Composed of inorganic compounds, such as oxides, carbides, nitrides, or metals silicates.	Composed of metallic elements or alloys.	Composed of two or more different materials with distinct properties.
Importance	They are biocompatible, biodegradable, and can mimic the structure and function of native tissues.	They offer more control over the design and fabrication of biomaterials with desired properties and functions.	They are bioactive, biostable, and resistant to corrosion and wear.	They are biostable, conductive, and able to withstand high loads and stresses.	They combine the advantages of different materials to achieve improved performance and functionality.
Quality	They vary in their chemical and physical properties depending on the source and processing methods.	They can be tailored to have specific molecular weight, structure, degradation rate, and surface properties.	They have high hardness, stiffness, and melting point.	They have high density, ductility, and malleability.	They can be engineered to have specific properties depending on composition and arrangement of the constituent materials.
Strength	They are generally weaker than synthetic polymers and may degrade over time.	They can have higher mechanical strength and stability than natural polymers and can be modified to enhance biocompatibility.	They have high compressive strength but low tensile strength, and fracture toughness.	They have high tensile strength and fatigue resistance but may corrode or release ions in biological environments.	They can have higher strength to weight ratio than individual materials and can be tailored to match the mechanical properties of native tissues.
Applications	They are used for tissue engineering, drug delivery, wound healing, and biosensors.	They are used for implants, scaffolds, stents, catheters, artificial organs, and contact lenses.	They are used for bone grafts, dental implants, joint replacements, and coatings.	I used for bone fixation devices, orthopaedic implants, dental implant, pacemakers, and electrodes.	They are used for bone substitutes, dental composites, tissue scaffolds, and hybrid implants.

7.2.4 PRINCIPLES OF BIOPRINTING

The principle of 3D printing involves:

1) Modelling: The first step in 3D printing is to create a digital 3D model of the desired object with precise measurements and specifications by using the computer-aided design (CAD) software (Seol et al., 2014).
2) Printing: The designed desired 3D object is printed in a physical object by building it layer by layer with the help of a digital bioprinter. The printing material, which can be a large build volume, is fed into the printer in a raw form, such as filament or resin. The printer then melts, solidifies, or cures the material layer by layer, to create the final object. Bioprinter follows a set of instructions, known as a G-code, which allows exactly where to place each layer of material (Seol et al., 2014).
3) Finishing: 3D-bioprinted objects sometimes may require some finishing work such as removing support structures, sanding rough edges, or applying a coat of paint or other finish to improve the appearance or functionality of the object by hand or with additional equipment, depending on the complexity of the object to make it ready for use (Seol et al., 2014).

7.3 3D PRINTING TECHNOLOGY

The use of biomaterials and 3D printing is an exciting area of development that has the potential to transform the field of medicine and other medical applications. The following type of 3D printing technology in medical applications has been used (Kumar et al., 2021):

7.3.1 FUSED DEPOSITION MODELLING (FDM)

FDM is a method of creating 3D objects directly from a CAD model using a filament extrusion process. In the basic FDM process, a plastic filament is heated and extruded through a nozzle to form layers on a platform, following the instructions provided by the CAD model. The extruder head moves in the X-Y plane, while the platform moves in the Z-direction to achieve the desired layer thickness. The overall process of fabricating parts typically involves four stages. First, a CAD solid model is created and converted into a stereo lithography (STL) format, which approximates the model's geometry. The STL file is then prepared using FDM software, including tasks such as determining the part's orientation, slicing it into thin layers, selecting FDM parameters, and generating supports. Proper orientation is crucial to minimize or eliminate the need for support and enhance the surface finish. FDM parameters can include factors like raster width, build style, raster angle, air gap, nozzle tip size, and model temperature. The pre-processed file, including the supports and model, is typically saved as a Stratasys Machine Language (SML) file or CMB file and then transferred to the FDM machine for the actual part-building process (Hu & Qin, 2020). Kouhi et al. have introduced a method for designing and fabricating medical models using CAD and an FDM system, specifically the FDM3000, with the aim of applying them in mandibular reconstructive surgery (Kouhi et al., 2008). In vitro studies and mechanical testing were conducted on it and the results demonstrated that it was non-toxic and exhibited excellent cell growth.

Advantages: It is one of the most cost-effective 3D bioprinter technology, as it requires only a simple setup and inexpensive materials. It uses a broad range of thermoplastic materials, including ABS, PLA, nylon, PETG. It is a simple 3D-printing method, making it easy to learn and operate. FDM printers are widely available and can be used in a range of settings, from hobbyists to professional applications.

Disadvantages: FDM parts can be weaker than those made using other 3D printing methods due to the layer-by-layer nature of the process. The parts of it often have visible layer lines, which result in a rough surface finish. FDM parts have limited detail and precision compared to other 3D-printing methods.

Applications: It is commonly used for rapid prototyping and educational settings due to its low cost and ease of use. The ability to create a consistent porous structure through the manipulation of FDM process parameters and build styles has made the FDM process very appealing for the production of 3D scaffolds and biomaterials for tissue engineering applications. In the field of biomedical engineering, this system has been utilized by researchers to manufacture medical models and drug delivery systems. It is used to create a wide range of consumer goods including phone cases to toys and various tools like moulds, jigs, and fixtures.

7.3.2 STEREOLITHOGRAPHY (SLA)

SLA is a technique that utilizes vat-photopolymerization, employing a laser beam to create intricate 3D structures. It is known for its exceptional capabilities in fabricating a wide range of objects, from micron-sized needles to life-size organs (Choudhury et al., 2022). This is achieved through its high resolution, precision, accuracy, and speed. SLA utilizes photopolymerization materials that undergo solidification when exposed to (particularly UV) light. The size and shape of the desired object are controlled by selective exposure of the material to low-power light such as typical UV from 200 to 400 nm and visible from 400 to 700 which are generated from a He-Cd/Nd:YVO4 laser (Choudhury et al., 2022; Dhariwala et al., 2004).

Advantages: It has the potential to revolutionize the traditional "one dose fits all" approach, enabling the realization of personalized, digitally driven patient-centric medication. It uses a wide range of material, including resins, flexible and heat-resistant biocompatible dispersion agents, drugs, photopolymers, photoinitiators, and plasticizers. It is known for creating highly precise and detailed objects with a smooth surface finish, making it ideal for creating objects with intricate geometries. It is a relatively fast 3D-printing method, suitable for producing small batches of parts quickly. It produces minimal waste as the excess resin can be reused.

Disadvantages: The key disadvantages are the high cost of printers and materials and the smaller build volume. SLA parts may be more brittle than those produced by other printing methods.

Applications: This method is commonly used in dentistry to create dental models, aligners, and other dental appliances. This technology has been utilized to fabricate different types of topical and transdermal delivery systems, including films and microneedles. For example, researchers have employed it to create a polymeric film for the delivery of berberine (BBR) by modifying the profiles of drug release and monitoring the tablet's geometrical features like in-fill density, volume ratio, surface area, etc.. It is also used to create customised medical devices such as surgical guides and implants.

7.3.3 Selective Laser Sintering (SLS)

SLS is a bioprinting technique that utilizes laser energy to selectively heat powder particles, resulting in their fusion and subsequent solidification to form a physical 3D structure. It consists of (1) a spreading platform, (2) a powder bed, and (3) a laser system with a scanner (Odde & Renn, 1999b, 1999a). To initiate the procedure, a spreading system is employed to uniformly distribute powder onto the building platform. It consists of a slot feeder and a roller/scraper blade, which work together to ensure an even surface. The 3D component is then processed in sequential layers, each represented by vectors that serve as the fundamental elements for laser scanning. During the process, the powder material is heated to a temperature below its melting point, but enough to induce fusion through laser sintering or melting between particles. The scanner moves the laser across a two-dimensional plane, while simultaneously adjusting the height of the powder bed to focus the laser on the newly formed surface. Loose powder particles on the build platform function as support during the procedure (Odde & Renn, 1999b). Once each layer is fused, the surface of the powder bed is lowered by the thickness of one layer, and a fresh layer of powder is added and fused using the laser. This layer-by-layer approach continues until the entire object is constructed. Upon completion of the printing process, the finished product is allowed to cool down within the printer. Finally, the product is extracted from the loose powder, either manually or through sieving.

Advantages: It also uses a materials wide range including plastics, metals, and ceramics, allowing for a diverse range of applications. SLS parts can be stronger than those produced using other methods due to the use of powdered materials. SLS parts are self-supporting during printing, meaning that no additional support structures are needed. It produces accurate and precise parts with a smooth surface finish.

Disadvantages: The printers and materials are expensive, and the resolution print is restricted by the size of the powder particles, resulting in lower resolution of the bioprinted product. SLS parts may have limited detail and precision compared to other methods.

Applications: It is commonly used in aerospace to create complex, high-strength components. It is also used for manufacturing medical devices, such as orthopaedic implants. SLS is a popular method for the rapid prototyping of complex parts and is used for small-batch production of customized parts, as well as for creating moulds for injection moulding.

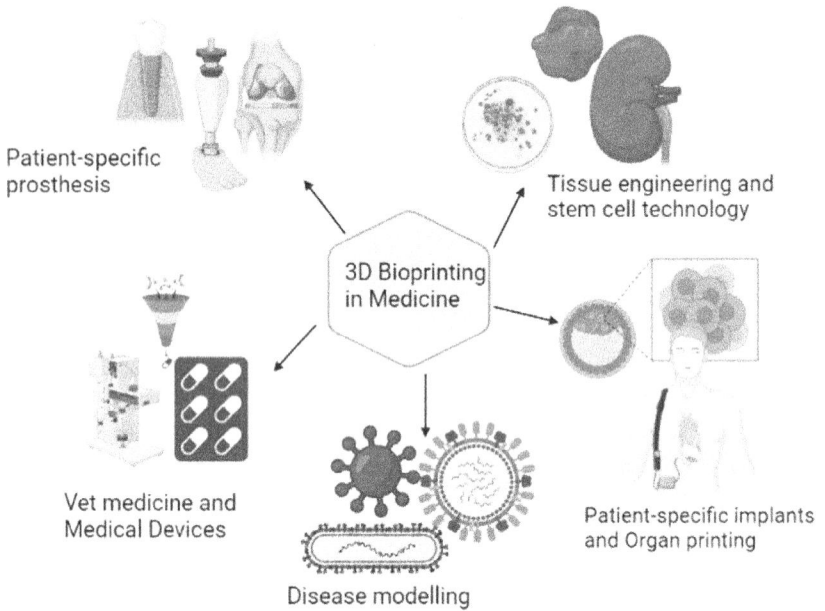

FIGURE 7.3 Medical applications of 3D printing technology.

7.3.4 BINDER JETTING

Binder Jetting involves depositing a liquid binder onto a powder bed, layer by layer, to create the final object. The powder is then cured, typically using heat, to solidify the object. In the field of bone tissue engineering, calcium phosphate (CaP) powders, such as tricalcium phosphate (TCP) or hydroxyapatite (HA), are commonly used. These powders serve as the main component for creating scaffolds or structures that support the regeneration of bone tissue. The binding solution used in the manufacturing process either can be a sacrificial polymer, which will be pyrolyzed or burned off during the sintering process after printing, or an aqueous solution, which is typically a dilute (concentration range of 5–30 wt.%) phosphoric acid solution. The sacrificial polymer acts as a temporary binder, holding the CaP particles (desired biomaterials) together until they can be fused during sintering. Once the polymer is burned off, the remaining CaP particles fuse together, creating a solid structure. Alternatively, the acidic nature of the aqueous solution of the binder starts a dissolution-precipitation reaction within the powder of CaP. This reaction causes the CaP particles to dissolve partially and then re-precipitate, fusing the particles together and forming a cohesive structure. The use of an aqueous binding solution is especially suitable for CaP powders because they readily react with phosphoric acid, facilitating particle fusion. The resulting structure, formed by either the sacrificial polymer or the aqueous binding solution, provides the necessary mechanical support for cell attachment, proliferation, and tissue regeneration. These scaffolds mimic the natural bone structure and gradually degrade over time, allowing new bone tissue to replace the scaffold

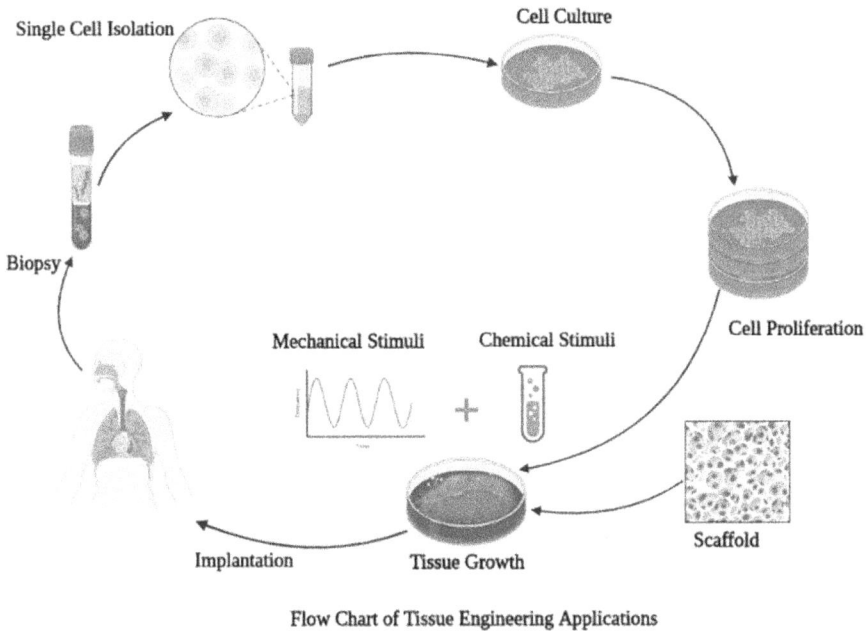

Flow Chart of Tissue Engineering Applications

FIGURE 7.4   Flow chart of tissue engineering.

material. It is important to note that the specific materials and techniques used in bone tissue engineering can vary depending on the desired application, research institution, or medical facility (Figure 7.4) (Choudhury et al., 2022).

Advantages: It is also a fast method of 3D printing; resultant small batches of parts are produced quickly. Like SLS, it also eliminates the need for supporting structures and the physical support of the powder bed itself allows for the production of complex geometries. This simplifies the manufacturing process and reduces post-processing requirements. It is allowed to reuse unused powders that can be collected, sieved, and reused in subsequent printing cycles. This helps to reduce material waste and improve cost-efficiency. Initially, water-based binders and gypsum-based powders were commonly used, however, nowadays versatile materials and binders are used. This technique also produces coloured parts by incorporating colourants into the binder. It has a large build volume that helps in the creation of larger objects.

Disadvantages: Binder Jetting parts may be weaker than parts produced using other methods of 3D printing due to the use of a binding agent. The resolution of a binder jetting print is limited by the size of the powder particles, which can result in lower resolution than other methods of 3D printing. Binder Jetting parts often require post-processing to remove excess powder and strengthen the parts.

Applications: It is often used for creating large parts or for printing with materials that are difficult to extrude. It can be used to create architectural models and components quickly and cost-effectively. Binder Jetting can be used to create complex and intricate jewellery designs. Binder Jetting can be used to create parts for the automotive industry, such as engine components and brackets. Binder Jetting can be used to create medical devices such as prosthetics, bone tissue engineering, and dental implants.

7.3.5 MATERIAL JETTING

It is a type of additive manufacturing process that uses inkjet-style printheads to deposit droplets of liquid material onto a build platform and cure them with light or heat to form a solid 3D object. The typical process is as follows (Gülcan et al., 2021): At first, optimal viscosity of liquid material by heating is achieved for printing. Then, the printhead moves over the build stage and hundreds of minute droplets of the desired material are jetted/deposited to the programmed locations. The printhead can use either a thermal or a piezoelectric method to eject the droplets. Next, the droplets of material are cured by either ultraviolet light or heat to form solid layers. The curing happens immediately after each layer is jetted. The process is repeated until the 3D object is completed. A fly cutter skims the build area after each layer to ensure a perfectly flat surface (Gülcan et al., 2021). Finally, the support material, which is also jetted during the build, is removed by hand or by water jetting. The support material is usually a gel-like substance that can support complex geometries. Some machines also apply to give colour to the model.

There are three main subtypes of material jetting (Elkaseer et al., 2022): (1) PolyJet: This technology jets ultra-thin layers of photopolymer material and cures them with UV light. It can also use gel-like support material for complex geometries. (2) Nano Particle Jetting (NPJ): This technology jets liquid containing nanoparticles of metal or ceramic and cures them with high temperatures. It can produce high-resolution parts with excellent mechanical properties. (3) Drop-On Demand (DOD): This technology jets wax-like material and cures it with heat. It is mainly used for creating patterns for investment casting or mould making.

Advantages: It produces parts with high resolution, detail, and smooth surface finish; resultant suitable for creating models with fine features and intricate designs. Like other 3D bioprinting it also offers to be used in a wide range of materials like photopolymers, waxes, ceramics, and metals. It can print multiple materials in a single part, allowing for the creation of parts with varying properties.

Disadvantages: The bioprinted product may be weaker than the other methods due to the use of photopolymer materials. Material Jetting parts often require post-processing to remove support structures and strengthen the parts and the cost of printers and materials is also very high.

Applications: It is a popular method for the rapid prototyping of complex parts and models. It is used to create dental and surgical models, prosthetics, and hearing aids. It is also used to create high-resolution jewellery designs and architectural models with fine details.

7.4   APPLICATIONS OF 3D PRINTING IN BIOMEDICAL AREAS

Numerous studies have recently found the use of 3D printing in biomedical applications to be intriguing, and many businesses around the world have contributed to this development with their research facilities and testing facilities. North America, with the United States and Canada, accounted for the highest number (combined total of 113 facilities) of centralized 3D printing facilities in hospitals and healthcare. While 48 facilities in Europe were equipped till 2019, they also recognized the potential of 3D printing in improving patient care and surgical outcomes (Kalita et al., 2003). Several countries in Europe, including Germany, the United Kingdom, the Netherlands, and Spain, have been active in implementing medical 3D-printing solutions. Popularity of Medical 3D printing is increasing worldwide due to its numerous benefits in various medical applications including surgical planning, personalized implants, anatomical models, and medical education (Thomas & Singh, 2020). Figure 7.3 represents some medical applications of 3D printing technology. The following are some key examples of its utility:

7.4.1 DISEASE MODELLING

Three-dimensional printing has been increasingly used in disease modelling to create patient-specific models of diseased tissues or organs for research and diagnostic purposes, for example, creating of patient's specific cancer and heart models to help researchers in the study and diagnosing of such cancer growth and drug resistance, identifying potential therapies, congenital heart defects, coronary artery disease, heart failure, etc. (Kačarević et al., 2018; Shahrubudin et al., 2020). Similarly in neurological disorders and orthopaedics, it is used to create models of the brain or brain tumours and bones and joints to help surgeons plan for surgeries, optimize treatment options, plan and simulate surgeries, and develop custom implants. Furthermore, in respiratory and infectious diseases such as asthma, chronic obstructive pulmonary disease, and lung cancer, it is used to create models of the respiratory system and pathogens for studying and helping researchers develop better treatment options. .

7.4.2 SURGICAL APPLICATIONS

Advanced imaging technology in 3D printing has improved surgical planning and provides surgeons with a better view of the patient's anatomical structure. This technology is particularly useful in guiding and supervising surgery procedures, estimate angles, bone direction, and size, especially in cardiovascular, neurosurgery, orthopaedic, plastic and aesthetic surgery, where it can improve surgical outcomes and reduce medical errors, increasing patient safety (Zein et al., 2013). Here are some case studies: (1) Heart: Yang et al. (2015) developed a 3D-printed heart using diverse colours and variable softness, which eased the documentation of different portions in the medical preparation. The heart model, with its different coloured zones, served as an excellent visualization tool, and the use of Stratasys' "Tango" materials family ensured the required softness. (2) Kidney: In the work conducted by Kusaka et al., a 3D-printed kidney was produced, employing the ability to print different parts with varying colours and hardness. This approach aimed to create models that closely mimicked the characteristics of soft living tissues, surpassing the limitations of many traditional mono-material 3D-printing techniques (Kusaka et al., 2015). Based on these case studies, it is widely accepted that this technology is useful in surgical applications and that its use may continue to increase in the coming years with further research and development (Krauel et al., 2016).

7.4.3 MEDICAL DEVICES

Three-dimensional printer technology is one of the popular methods for producing medical devices, especially those that are difficult to manufacture using traditional methods. It offers affordable production of devices that are suitable for the patient's anatomical structure, for instance, as yet 10 million hearing aids devices have been printed worldwide as reported in most reports (*How 3D Printed Hearing Aids Silently Took Over The World – 3DSourced*, n.d.). However, SONOVA itself claimed that they have produced 1 million hearing aids in the year 2016 (*3D Printing Technology for Improved Hearing | Sonova International*, n.d.; *How 3D Printed Hearing Aids Silently Took Over The World – 3DSourced*, n.d.). In addition to hearing aids, this technology shows promise for other medical devices like eye lenses, stethoscopes, and glasses for the visually impaired. The use of 3D printing is expected to increase in the future, given its versatility in producing a broad range of medical devices from prostheses to human organs. It can also serve as an alternative to clinical animals and human trails in toxicology tests. Finally, the use of 3D printers offers designing of custom-desired geometries and dimensions of drugs, resultant cost and time of the drug development with complex formulations will be reduced.

7.4.4 PATIENT-SPECIFIC IMPLANTS

The use of 3D-printing technology for manufacturing personalized prostheses for patients is crucial in the medical field. It offers benefits such as convenience for healthcare providers and increased patient comfort. Personalised implants are produced using image scanning and 3D printing, which eliminates the need for traditional measurement methods and reduces waiting times. This method is commonly used for knee and dental implants, tibial, femoral, fibular, and acetabular implants (Lee et al., 2020; Moret et al., 2021). It also supports tissue regeneration and has high fatigue strength and corrosion resistance. Cranial implants are one example of successful 3D-printed implants that can be customized to a patient's anatomy, improve appearance, and provide psychological support. The Medical Design and Production Center (METUM) in Turkey is an organization that has successfully produced cranial implants using 3D printing for patients such as Yasar Agayev, who was shot in the head in April 2016 (Bozkurt & Karayel, 2021). Thanks to this technology, METUM and other organizations around the world can create personalized implants with good mechanical properties that are suitable for patients.

7.4.5 PATIENT-SPECIFIC PROSTHESIS

Three-dimensional printing technology has emerged as a successful alternative in recent years of ancient Egyptian prostheses development through casting (Crimì et al., 2023). Compared to traditional methods, this technology allows more realistic prostheses production, compatible with the patient's anatomy, and has the desired mechanical and physical properties. Prostheses can be produced for various body parts, such as the ear, nose, teeth, bone, hand, and foot, and can be adjusted to match the skin tone of the patient (Sherwood et al., 2020). Prostheses can be also produced in different colours to accommodate people of various ages, races, genders, and sizes (Thomas & Singh, 2020). The 3D-printed models helped us to plan the surgical approach, select the appropriate prosthesis size and type, and simulate the surgery before the actual operation. The models also served as educational tools for the patients and trainees. (Thomas & Singh, 2020). Here is a case of a 65-year-old female patient who had severe osteoarthritis of the left hip with acetabular dysplasia and protrusio acetabuli. The 3D-printed model of her hip joint showed a large medial wall defect and a shallow acetabulum that required a custom-made porous metal augment and a large-diameter ceramic-on-ceramic bearing. The model also helped to determine the optimal position and orientation of the acetabular component and the femoral stem. The surgery was performed successfully with minimal blood loss and no intraoperative complications. The patient reported excellent pain relief and functional improvement at 6 months follow-up (Willemsen et al., 2019).

7.4.6 VET MEDICINE

In veterinary medicine, 3D printing is often used for pre-surgery planning and imaging, which can help reduce errors and improve the safety of the animal during surgery. It has also been used successfully in various surgeries for animals, such as tumour removal and the production of prostheses for injured or disabled animals. Additionally, the technology offers to create complex implants. One example of 3D printing in veterinary medicine is the use of patient-specific drill guides for pedicle screw placement in spinal surgery. Pedicle screws are used to stabilize the spine and correct deformities or fractures. However, placing them accurately and safely can be challenging, especially in small animals with complex anatomy (Gyles, 2019). A case study of the University of Zurich and ETH Zurich developed a method to design and print patient-specific drill guides for pedicle screw placement in dogs. They used computed tomography (CT) scans to create 3D models of the spine and the screws, and then used a piece of software to design the drill guides. The guides were printed using a stereolithography (SLA) printer with a biocompatible resin (Gyles, 2019). The results showed that the 3D-printed drill guides improved the accuracy and safety of pedicle screw placement compared to conventional methods. The guides also reduced the operative time and radiation exposure (Dautzenberg et al., 2021; Gyles, 2019).

7.4.7 TISSUE ENGINEERING AND STEM CELL TECHNOLOGY

The fields of tissue engineering and stem cell technology are both being advanced by the use of 3D printing. This is achieved by producing artificial tissues and organs in a laboratory setting, using living cells on scaffolds using engineering and life sciences principles together (Mani et al., 2022). It is being used to produce stem cells that can be transformed into blood or muscle cells, adipose tissues, and even blood vessels, which have the potential to replace damaged tissues. A few years back, Wyss Institute and Harvard SEAS developed a novel 3D-bioprinting technique that can generate living human cells composed of vascularized tissues which are approximately ten times thicker than any previously bioengineered tissues yet (*3D Bioprinting of Living Tissues*, n.d.). Using a multi-material 3D printer, they created a silicone mould to house and connect the printed tissue on a chip. Then, they printed large vascular channels containing endothelial cells in silicone ink within the mould, forming an artificial vasculature (*3D Bioprinting of Living Tissues*, n.d.). Additionally, a self-supporting ink containing mesenchymal stem cells (MSCs) was printed separately to form the tissue parenchyma. The remaining open regions were filled with a liquid containing fibroblasts and extracellular matrix, contributing a connective tissue element. This resulted in a soft tissue construct that could be immediately perfused with nutrients and growth factors through a single inlet and outlet, connected to the vascular channel. The construct maintained its structure and function for at least six weeks (*3D Bioprinting of Living Tissues*, n.d.).

7.4.8 ORGAN PRINTING

Bioprinting has been used to create various tissues, such as multilayer heart tissue, cartilage, and bone structures. Despite the complex structure of the ear, a bionic ear was produced using bioprinting by anatomical geometry (Dey & Ozbolat, 2020). Wake Forest Institute for Regenerative Medicine achieved a breakthrough in organ printing by successfully implanting 3D bioprinted kidney bladders into three children who suffered from bladder dysfunction due to spina bifida (Dey & Ozbolat, 2020). They used a combination of synthetic and natural materials to create a scaffold that matched the shape and size of the patient's bladder. Then, they seeded the scaffold with the patient's own bladder cells and smooth muscle cells, which were obtained by biopsy. The cells were cultured on the scaffold for about seven weeks before implantation (Dey & Ozbolat, 2020). The results showed that the 3D-bioprinted bladders improved the bladder function and quality of life of the patients. The bladders also exhibited vascularization and innervation after implantation (Dey & Ozbolat, 2020).

7.5 CURRENT CHALLENGES

Despite the potential of bioprinting in various medical applications, there are several obstacles to overcome. For example, it should be compatible and able to withstand the high temperatures and pressures involved in the printing process, without losing its properties or causing damage to the printer (Mao et al., 2020). In the case of 3D-printed prosthesis, the hardness of it compared to the original skin and not being able to adjust as the patient grows, will be a huge restriction on its uses. Additionally, the weight of prostheses can be a problem, especially for complex prostheses, which can make daily activities difficult. Despite the vast potential of bioprinting in regenerative medicine, there are several challenges, such as material and cell requirements, tissue maturation, and functionality. It also includes differences between individuals that present a significant challenge, as each patient responds differently to existing drug treatments. One of the biggest challenges of 3D bioprinting is the assembling of vascular structures, which is a problem for most tissue engineering technologies (Kačarević et al., 2018). 3D-printed organs or tissues without a circulatory system face nutrient and waste removal problems, leading to cell viability issues and organ malfunction. Addressing these challenges will require continued research and development, as well as collaboration across disciplines.

7.6 FUTURE PROSPECTIVE

Three-dimensional printing technology, which has proven to be successful in various fields, offers promising opportunities for the future and may help solve some of the problems we face today such as with the help of CAD-CAM programs, any type of mask can be quickly modelled and printed (Kumar et al., 2021). This technology holds promise in the field of cancer research, which is a major health problem worldwide and the second-leading cause of death. Chemotherapy side effects like vomiting and heart failure can be reduced by developing new biomaterials that can be used to deliver drugs directly to specific parts of the body and reducing the amount of drug administered into the body with a 3D sponge project. This technology also has the potential to treat congenital disorders or birth defects, such as cleft palates, and is used in the education of medical students. Additionally, they can be used to replace or repair damaged tissue, to create scaffolds for tissue engineering, deliver drugs to specific areas of the body, and provide scaffolding for the growth of new tissue (Bakhtiar et al., 2018). In future, it may even be possible to print 3D copies of organs that mimic the properties of living human tissue, which could have significant implications for drug development, treatment testing, medical research, wound healing, and organ transplantation. This technology also prevents waste by manufacturing on demand. In conclusion, the integration of 3D printing and biomaterials holds significant potential for advancing medical treatments and therapies, paving the way for a higher quality of life and increased lifespan for patients.

7.7 CONCLUSION

The following conclusion can be drawn from this chapter:

- Biomaterials and 3D printing have revolutionized the field of biomedical applications, allowing for the creation of customized implants, tissues, and medical equipment.
- 3D printing offers advantages over traditional manufacturing techniques, such as the ability to produce complex and customized geometries with high accuracy and reproducibility.
- Various materials, including polymers, ceramics, metals, and composites, can be used for 3D printing in biomedical applications, each with its own properties and processing requirements.

- Different 3D printing techniques, such as fused deposition modelling (FDM), stereolithography (SLA), and selective laser sintering (SLS), have their own advantages and limitations.
- The combination of precision 3D printing and versatile biomaterials has the potential to improve outcomes for patients worldwide through personalized, customized, and regenerative therapies.
- It has various applications in the biomedical field, including surgical applications, medical device manufacturing, tissue engineering, and organ bioprinting.
- While not yet widespread, the use of 3D printing in the pharmaceutical sector and bioprinting applications is expected to bring revolutionary developments in the medical field, from cancer treatment to functional prostheses and longer lifespans for patients.

REFERENCES

Bakhtiar, S. M., Butt, H. A., Zeb, S., Quddusi, D. M., Gul, S., & Dilshad, E. (2018). 3D printing technologies and their applications in biomedical science. In *Omics Technologies and Bio-Engineering* (pp. 167–189). Academic Press. https://doi.org/10.1016/B978-0-12-804659-3.00010-5

Bozkurt, Y., & Karayel, E. (2021). 3D printing technology; methods, biomedical applications, future opportunities and trends. *Journal of Materials Research and Technology*, *14*, 1430–1450. https://doi.org/10.1016/J.JMRT.2021.07.050

Choudhury, D., Sharma, P. K., Suryanarayana Murty, U., & Banerjee, S. (2022). Stereolithography-assisted fabrication of 3D printed polymeric film for topical berberine delivery: In-vitro, ex-vivo and in-vivo investigations. *Journal of Pharmacy and Pharmacology*, *74*(10), 1477–1488. https://doi.org/10.1093/JPP/RGAB158

Crimì, A., Joyce, D. M., Binitie, O., Ruggieri, P., & Letson, G. D. (2023). The history of resection prosthesis. *International Orthopaedics*, *47*(3), 873–883. https://doi.org/10.1007/S00264-023-05698-W/METRICS

Dababneh, A. B., & Ozbolat, I. T. (2014). Bioprinting technology: A current state-of-the-art review. *Journal of Manufacturing Science and Engineering*, *136*(6), 061016. https://doi.org/10.1115/1.4028512/377608

Dautzenberg, P., Volk, H. A., Huels, N., Cieciora, L., Dohmen, K., Lüpke, M., … Harms, O. (2021). The effect of steam sterilization on different 3D printable materials for surgical use in veterinary medicine. *BMC Veterinary Research*, *17*(1), 1–6. https://doi.org/10.1186/S12917-021-03065-8/FIGURES/3

De Santis, M. M., Alsafadi, H. N., Tas, S., Bölükbas, D. A., Prithiviraj, S., Da Silva, I. A. N., Mittendorfer, M., Ota, C., Stegmayr, J., Daoud, F., Königshoff, M., Swärd, K., Wood, J. A., Tassieri, M., Bourgine, P. E., Lindstedt, S., Mohlin, S., & Wagner, D. E. (2021). Extracellular-matrix-reinforced bioinks for 3D bioprinting human tissue. *Advanced Materials*, *33*(3). https://doi.org/10.1002/ADMA.202005476

Dey, M., & Ozbolat, I. T. (2020). 3D bioprinting of cells, tissues and organs. *Scientific Reports*, *10*(1), 14023. https://doi.org/10.1038/s41598-020-70086-y

Dhariwala, B., Hunt, E., & Boland, T. (2004). Rapid prototyping of tissue-engineering constructs, using photopolymerizable hydrogels and stereolithography. *Tissue Engineering*, *10*(9–10), 1316–1322. https://doi.org/10.1089/TEN.2004.10.1316

Donderwinkel, I., Van Hest, J. C., & Cameron, N. R. (2017). Bio-inks for 3D bioprinting: Recent advances and future prospects. *Polymer Chemistry*, *8*(31), 4451–4471. https://doi.org/10.1039/C7PY00826K

Duan, B. (2017). State-of-the-art review of 3D bioprinting for cardiovascular tissue engineering. *Annals of Biomedical Engineering*, *45*, 195–209. https://doi.org/10.1007/S10439-016-1607-5

Elkaseer, A., Chen, K. J., Janhsen, J. C., Refle, O., Hagenmeyer, V., & Scholz, S. G. (2022). Material jetting for advanced applications: A state-of-the-art review, gaps and future directions. *Additive Manufacturing*, 103270. https://doi.org/10.1016/J.ADDMA.2022.103270

Foty, R. A., Pfleger, C. M., Forgacs, G., & Steinberg, M. S. (1996). Surface tensions of embryonic tissues predict their mutual envelopment behavior. *Development*, *122*(5), 1611–1620. https://doi.org/10.1242/DEV.122.5.1611

Gao, Q., He, Y., Fu, J. Z., Liu, A., & Ma, L. (2015). Coaxial nozzle-assisted 3D bioprinting with built-in microchannels for nutrients delivery. *Biomaterials*, *61*, 203–215. https://doi.org/10.1016/J.BIOMATERIALS.2015.05.031

Garcia-Cruz, M. R., Postma, A., Frith, J. E., & Meagher, L. (2021). Printability and bio-functionality of a shear thinning methacrylated xanthan–gelatin composite bioink. *Biofabrication*, *13*(3), 035023. https://doi.org/10.1088/1758-5090/ABEC2D

Gülcan, O., Günaydın, K., & Tamer, A. (2021). The state of the art of material jetting—A critical review. *Polymers*, *13*(16), 2829. https://doi.org/10.3390/POLYM13162829

Gyles, C. (2019). 3D printing comes to veterinary medicine. *The Canadian Veterinary Journal*, *60*(10), 1033. /pmc/articles/PMC6741827/

Hu, C., & Qin, Q. H. (2020). Advances in fused deposition modeling of discontinuous fiber/polymer composites. *Current Opinion in Solid State and Materials Science*, *24*(5), 100867. https://doi.org/10.1016/J.COSSMS.2020.100867

Jayasinghe, S. N., Qureshi, A. N., & Eagles, P. A. (2006). Electrospraying living cells. *Small*, *2*(2), 216–219. https://doi.org/10.1002/SMLL.200500291

Kačarević, Ž. P., Rider, P. M., Alkildani, S., Retnasingh, S., Smeets, R., Jung, O., & Barbeck, M. (2018). An introduction to 3D bioprinting: Possibilities, challenges and future aspects. *Materials*, *11*(11), 2199. https://doi.org/10.3390/MA11112199

Kalita, S. J., Bose, S., Hosick, H. L., & Bandyopadhyay, A. (2003). Development of controlled porosity polymer-ceramic composite scaffolds via fused deposition modeling. *Materials Science and Engineering: C*, *23*(5), 611–620. https://doi.org/10.1016/S0928-4931(03)00052-3

Kalman, H., Tripathi, N. M., Gabrieli, O. G., & Portnikov, D. (2017). Phase diagrams for pneumatic and hydraulic conveying. In *18th International Conferences on Transport and Sedimentation of Solid Particles, T and S 2017* (pp. 145–152). Wydawnictwo Uniwersytetu Przyrodniczego we Wroclawiu.

Kang, H. W., Lee, S. J., Ko, I. K., Kengla, C., Yoo, J. J., & Atala, A. (2016). A 3D bioprinting system to produce human-scale tissue constructs with structural integrity. *Nature Biotechnology*, *34*(3), 312–319. https://doi.org/10.1038/NBT.3413

Karzyński, K., Kosowska, K., Ambroźkiewicz, F., Berman, A., Cichoń, J., Klak, M., Serwańska-Świętek, M., & Wszoła, M. (2018). Use of 3D bioprinting in biomedical engineering for clinical application. *Medical Studies/Studia Medyczne*, *34*(1), 93–97. https://doi.org/10.5114/MS.2018.74827

Klebe, R. J. (1988). Cytoscribing: A method for micropositioning cells and the construction of two- and three-dimensional synthetic tissues. *Experimental Cell Research*, *179*(2), 362–373. https://doi.org/10.1016/0014-4827(88)90275-3

Kouhi, E., Masood, S., & Morsi, Y. (2008). Design and fabrication of reconstructive mandibular models using fused deposition modeling. *Assembly Automation*, *28*(3), 246–254. https://doi.org/10.1108/01445150810889501/FULL/XML

Krauel, L., Fenollosa, F., Riaza, L., Pérez, M., Tarrado, X., Morales, A., Gomà, J., & Mora, J. (2016). Use of 3D prototypes for complex surgical oncologic cases. *World Journal of Surgery*, *40*(4), 889–894. https://doi.org/10.1007/S00268-015-3295-Y/METRICS

Kumar, P., Rajak, D. K., Abubakar, M., Ali, S. G. M., & Hussain, M. (2021). 3D printing technology for biomedical practice: A review. *Journal of Materials Engineering and Performance*, *30*(7), 5342–5355. https://doi.org/10.1007/S11665-021-05792-3/METRICS

Kusaka, M., Sugimoto, M., Fukami, N., Sasaki, H., Takenaka, M., Anraku, T., Ito, T., Kenmochi, T., Shiroki, R., & Hoshinaga, K. (2015). Initial experience with a tailor-made simulation and navigation program using a 3-D printer model of kidney transplantation surgery. *Transplantation Proceedings*, *47*(3), 596–599. https://doi.org/10.1016/J.TRANSPROCEED.2014.12.045

Landers, R., Hübner, U., Schmelzeisen, R., & Mülhaupt, R. (2002). Rapid prototyping of scaffolds derived from thermoreversible hydrogels and tailored for applications in tissue engineering. *Biomaterials*, *23*(23), 4437–4447. https://doi.org/10.1016/S0142-9612(02)00139-4

Lee, J. A., Koh, Y. G., & Kang, K. T. (2020). Biomechanical and clinical effect of patient-specific or customized knee implants: A review. *Journal of Clinical Medicine*, *9*(5), 1559. https://doi.org/10.3390/JCM9051559

Liu, N., Zhang, X., Guo, Q., Wu, T., & Wang, Y. (2022). 3D bioprinted scaffolds for tissue repair and regeneration. *Frontiers in Materials*, *9*, 445. https://doi.org/10.3389/FMATS.2022.925321/BIBTEX

Lumay, G., Tripathi, N. M., & Francqui, F. (2019). How to gain a full understanding of powder flow properties, and the benefits of doing so. *ONdrugDelivery*, *2019*, 42–47.

Mani, M. P., Sadia, M., Jaganathan, S. K., Khudzari, A. Z., Supriyanto, E., Saidin, S., & Faudzi, A. A. M. (2022). A review on 3D printing in tissue engineering applications. *Journal of Polymer Engineering*, *42*(3), 243–265. https://doi.org/10.1515/POLYENG-2021-0059/ASSET/GRAPHIC/J_POLYENG-2021-0059_FIG_004.JPG

Mao, H., Yang, L., Zhu, H., Wu, L., Ji, P., Yang, J., & Gu, Z. (2020). Recent advances and challenges in materials for 3D bioprinting. *Progress in Natural Science: Materials International*, *30*(5), 618–634. https://doi.org/10.1016/J.PNSC.2020.09.015

Moret, C. S., Schelker, B. L., & Hirschmann, M. T. (2021). Clinical and radiological outcomes after knee arthroplasty with patient-specific versus off-the-shelf knee implants: A systematic review. *Journal of Personalized Medicine*, *11*(7), 590. https://doi.org/10.3390/JPM11070590

M Tripathi, N., & S Mallick, S. (2017). Pneumatic conveying of Fly Ash: Bend Models investigation. *Advanced Materials Proceedings*, *2*(8), 526–531.

Naghieh, S., & Chen, X. (2021). Printability–A key issue in extrusion-based bioprinting. *Journal of Pharmaceutical Analysis*, *11*(5), 564–579. https://doi.org/10.1016/J.JPHA.2021.02.001

Neveu, A., Lumay, G., Pillitteri, S., Monsuur, F., Pauly, T., Ribeyre, Q., Francqui, F., Vandewalle, N., & Tripathi, N. M. (2020b). Physical characterization of blends containing mesoporous particles with a focus on electrostatic properties. In *2020 AIChE Spring Meeting and 16th Global Congress on Process Safety*. AIChE.

Neveu, A., Tripathi, N. M., Rigo, O., Francqui, F., & Lumay, G. (2020a). Experimental investigation of spreadability of metal powders in recoating process. In *Proceedings - Euro PM2020 Congress and Exhibition (Proceedings - Euro PM2020 Congress and Exhibition)*. European Powder Metallurgy Association (EPMA).

Ng, W. L., Lee, J. M., Yeong, W. Y., & Naing, M. W. (2017). Microvalve-based bioprinting–process, bio-inks and applications. *Biomaterials Science*, *5*(4), 632–647. https://doi.org/10.1039/c6bm00861e

Noor, N., Shapira, A., Edri, R., Gal, I., Wertheim, L., & Dvir, T. (2019). 3D printing of personalized thick and perfusable cardiac patches and hearts. *Advanced Science*, *6*(11), 1900344. https://doi.org/10.1002/ADVS.201900344

Norotte, C., Marga, F. S., Niklason, L. E., & Forgacs, G. (2009). Scaffold-free vascular tissue engineering using bioprinting. *Biomaterials, 30*(30), 5910–5917. https://doi.org/10.1016/J.BIOMATERIALS.2009.06.034

Odde, D. J., & Renn, M. J. (1999). Laser-guided direct writing for applications in biotechnology. *Trends in Biotechnology, 17*(10), 385–389. https://doi.org/10.1016/S0167-7799(99)01355-4

Ozbolat, I. T., Moncal, K. K., & Gudapati, H. (2017). Evaluation of bioprinter technologies. *Additive Manufacturing, 13*, 179–200. https://doi.org/10.1016/J.ADDMA.2016.10.003

Pyo, S. H., Wang, P., Hwang, H. H., Zhu, W., Warner, J., & Chen, S. (2017). Continuous optical 3D printing of green aliphatic polyurethanes. *ACS Applied Materials & Interfaces, 9*(1), 836–844. https://doi.org/10.1021/ACSAMI.6B12500

Ratsimba, A., Zerrouki, A., Tessier-Doyen, N., Nait-Ali, B., André, D., Duport, P., Neveu, A., Tripathi, N., Francqui, F., & Delaizir, G. (2021). Densification behaviour and three-dimensional printing of Y2O3 ceramic powder by selective laser sintering. *Ceramics International, 47*(6), 7465–7474. https://doi.org/10.1016/j.ceramint.2020.11.087

Seol, Y. J., Kang, H. W., Lee, S. J., Atala, A., & Yoo, J. J. (2014). Bioprinting technology and its applications. *European Journal of Cardio-Thoracic Surgery, 46*(3), 342–348. https://doi.org/10.1093/EJCTS/EZU148

Shahrubudin, N., Koshy, P., Alipal, J., Kadir, M. H. A., & Lee, T. C. (2020). Challenges of 3D printing technology for manufacturing biomedical products: A case study of Malaysian manufacturing firms. *Heliyon, 6*(4). https://doi.org/10.1016/J.HELIYON.2020.E03734

Sharma, A., Babbar, A., Tian, Y., Pathri, B. P., Gupta, M., & Singh, R. (2022). Machining of ceramic materials: A state of the art review. *International Journal on Interactive Design and Manufacturing (IJIDeM)*, 1–21. https://doi.org/10.1007/s12008-022-01016-7

Sharma, A., & Jain, V. (2020). Experimental investigation of cutting temperature during drilling of float glass specimen. *IOP Conference Series: Materials Science and Engineering* (Vol. 715, No. 1, p. 012050). IOP Publishing.

Sharma, A., Jain, V., & Gupta, D. (2018). Characterization of chipping and tool wear during drilling of float glass using rotary ultrasonic machining. *Measurement, 128*, 254–263.

Sharma, A., Jain, V., & Gupta, D. (2019a). Tool wear analysis while creating blind holes on float glass using conventional drilling: A multi-shaped tools study. In *Advances in Manufacturing Processes: Select Proceedings of ICEMMM 2018* (pp. 175–183). Springer Singapore.

Sharma, A., Jain, V., & Gupta, D. (2019b). Comparative analysis of chipping mechanics of float glass during rotary ultrasonic drilling and conventional drilling: For multi-shaped tools. *Machining Science and Technology, 23*(4), 547–568.

Sharma, A., Jain, V., & Gupta, D. (2019c). Multi-shaped tool wear study during rotary ultrasonic drilling and conventional drilling for amorphous solid. *Proceedings of the Institution of Mechanical Engineers, Part E: Journal of Process Mechanical Engineering, 233*(3), 551–560.

Sharma, A., Jain, V., & Gupta, D. (2021a). Effect of pre and post tempering on hole quality of float glass specimen: For rotary ultrasonic and conventional drilling. *Silicon, 13*, 2029–2039.

Sharma, A., Jain, V., & Gupta, D. (2021b). Mathematical approach on chipping volume estimation generated during rotary ultrasonic drilling for float glass. *Proceedings of the National Academy of Sciences, India Section A: Physical Sciences, 92*, 285–291.

Sharma, A., Jain, V., Gupta, D., & Babbar, A. (2020). A review study on miniaturization. In *Advanced Manufacturing and Processing Technology* (First edition, 111–131). CRC Press.

Sharma, A., Kalsia, M., Uppal, A. S., Babbar, A., & Dhawan, V. (2022). Machining of hard and brittle materials: A comprehensive review. *Materials Today: Proceedings, 50*, 1048–1052.

Sherwood, R. G., Murphy, N., Kearns, G., & Barry, C. (2020). The use of 3D printing technology in the creation of patient-specific facial prostheses. *Irish Journal of Medical Science (1971–)*, *189*, 1215–1221. https://doi.org/10.1007/S11845-020-02248-W

Skardal, A., Mack, D., Kapetanovic, E., Atala, A., Jackson, J. D., Yoo, J., & Soker, S. (2012). Bioprinted amniotic fluid-derived stem cells accelerate healing of large skin wounds. *Stem Cells Translational Medicine*, *1*(11), 792–802. https://doi.org/10.5966/SCTM.2012-0088

Tetsuka, H., & Shin, S. R. (2020). Materials and technical innovations in 3D printing in biomedical applications. *Journal of Materials Chemistry B*, *8*(15), 2930–2950. https://doi.org/10.1039/D0TB00034E

Thomas, D. J., & Singh, D. (2020). 3D printing for developing patient specific cosmetic prosthetics at the point of care. *International Journal of Surgery*, *80*, 241–242. https://doi.org/10.1016/J.IJSU.2020.04.023

Tripathi, N. M., Francqui, F., & Lumay, G. (2020). Influence of relative air humidity on the flow property of fine powders. In *Third International Conference on Powder, Granule and Bulk Solids: Innovations and Applications PGBSIA 2020 February 26–28, 2020* (p. 63).

Tripathi, N. M., Francqui, F., Pirenne, T., & Lumay, G. (2019). Measuring food powders electrical properties as a result of anti-static content. In *2019 AIChE Annual Meeting*. American Institute of Chemical Engineers.

Tripathi, N. M., Levy, A., & Kalman, H. (2016). Initial acceleration pressure drop in dilute phase pneumatic conveying system. In *Powder, Granule and Bulk Solids: Innovations and Applications Conference*.

Tripathi, N. M., & Mallick, S. S. (2014). *An Investigation into Pressure Drop Across Bends for Fluidised Densephase Pneumatic Conveying Systems* (Doctoral dissertation).

Willemsen, K., Nizak, R., Noordmans, H. J., Castelein, R. M., Weinans, H., & Kruyt, M. C. (2019). Challenges in the design and regulatory approval of 3D-printed surgical implants: A two-case series. *The Lancet Digital Health*, *1*(4), e163–e171. https://doi.org/10.1016/S2589-7500(19)30067-6

Wilson, W. C., & Boland, T. (2003). Cell and organ printing 1: protein and cell printers. *The Anatomical Record: Part A, Discoveries in Molecular, Cellular, and Evolutionary Biology*, *272*(2), 491–496. https://doi.org/10.1002/AR.A.10057

Yang, D. H., Kang, J. W., Kim, N., Song, J. K., Lee, J. W., & Lim, T. H. (2015). Myocardial 3-dimensional printing for septal myectomy guidance in a patient with obstructive hypertrophic cardiomyopathy. *Circulation*, *132*(4), 300–301. https://doi.org/10.1161/CIRCULATIONAHA.115.015842

Zein, N. N., Hanouneh, I. A., Bishop, P. D., Samaan, M., Eghtesad, B., Quintini, C., Miller, C., Yerian, L., & Klatte, R. (2013). Three-dimensional print of a liver for preoperative planning in living donor liver transplantation. *Liver Transplantation*, *19*(12), 1304–1310. https://doi.org/10.1002/LT.23729

Zhang, Y. S., Haghiashtiani, G., Hübscher, T., Kelly, D. J., Lee, J. M., Lutolf, M., McAlpine, M. C., Yeong, W. Y., Zenobi-Wong, M., & Malda, J. (2021). 3D extrusion bioprinting. *Nature Reviews Methods Primers*, *1*(1), 1–20. https://doi.org/10.1038/s43586-021-00073-8

8 Effect of Interlayer Temperatures and Heat Inputs on Porosity and Hydrogen Solubility in Wire Arc Additive Manufactured AA2618 Aluminium

A Comparative Study between Pulsed-MIG and CMT Methods

*Satishkumar P, Barun Haldar, Naveen Mani Tripathi,
Ankit Sharma, Dhaval Jaydevkumar Desai,
Vijay Kumar Sharma Seenivasan S, and Atul Babbar*

8.1 INTRODUCTION

WAAM garnered considerable interest from both researchers and industry experts due to its notable deposition speed, adaptability, and capability to produce parts unrestricted by size (Hussein et al. 2023). These attributes, combined with substantial material cost savings and applications in aerospace, made premium materials like Ti6Al-4V and Inconel more economically viable for WAAM compared to traditional processes. A significant focus was placed on the end product's microstructure, its mechanical strengths, and onsite finishing techniques (Gao et al. 2023). However, challenges arose from WAAM's unique microstructural attributes, its diminished mechanical performance in comparison to traditionally

DOI: 10.1201/9781003428862-8

processed items, and the handling of remaining stresses (Srinivasan et al. 2023; Manikandan et al. 2023).

Aluminium alloys have their affordability and extensive utilization in automotive and aerospace domains, and underwent rigorous examination (Xian et al. 2022). Apart from previously mentioned challenges, WAAM-produced aluminium components often have practical issues like porosity due to hydrogen incorporation and intergranular fractures. A disparity in the hydrogen solubility between solid (0.4 ml/kg) and liquid (7 ml/kg) aluminium contributed to this porosity issue (Aldalur, Suárez, & Veiga 2021; Wieczorowski et al. 2023; Ponomareva et al. 2021). The main sources of hydrogen included impurities on filler wire surfaces such as moisture, grease, and hydrocarbons (Satishkumar, Natarajan et al. 2022). Shielding gas, hoses, tubes, and substrates further intensified the hydrogen content. This hydrogen, upon interacting with contaminants, was quickly assimilated by the liquid aluminium as atomic hydrogen.

CMT emerged as a promising metal transfer technique, demonstrating efficacy in reducing porosity due to its distinctive deposition style and minimized heat input achieved through electronic and mechanical adjustments (Huang et al. 2023). Its low dilution levels and enhanced control over metal droplet transfer were explored for welding thin aluminium plates. When paired with interlayer rolling, CMT exhibited positive outcomes in minimizing porosity and fostering a commendable microstructure (Satishkumar et al. 2021). Whereas, several studies show an effective influence on additive manufactured materials by deploying advanced manufacturing technologies (Sharma et al. 2022; Sharma & Jain 2020; Sharma et al. 2018; Sharma et al. 2019a; Sharma et al. 2019b; Sharma et al. 2019c; Sharma et al. 2021a; Sharma et al. 2021b; Sharma et al. 2020; Sharma et al. 2022; Kalman et al. 2017; Lumay et al. 2019; Tripathi & Mallick 2017; Neveu et al. 2020a; Neveu et al. 2020b; Ratsimba et al. 2021; Tripathi et al. 2020; Tripathi et al. 2019; Tripathi et al. 2016; Tripathi and Mallick 2014).

During the sequential heating in metal layering, the deposited metal underwent reheating, influencing its microstructure, mechanical characteristics, and residual tensions (Han et al. 2020). Therefore, temperature regulation and heat management were deemed critical for optimal results in robotic metal deposition. Though interlayer waiting periods occasionally assisted in temperature control, they sometimes proved inadequate depending on the product's form and dimensions (Mayakannan et al. 2022). Adjusting the temperature of the last layer prior to the addition of the next one appeared to be an effective solution. Researchers (Satishkumar, Saravana Murthi, et al. 2022) employed a similar method, preheating the base material within specific temperature brackets for the initial and following layers. Adhering to the BS EN 1011-4:2000 guidelines, which advised a peak inter-pass temperature of 130°C for AA 6xxx weld materials, yielded positive outcomes (Ghoncheh et al. 2020).

This study explored the impact of heat input, temperature between layers, and time lapse between layers on porosity during the deposition process. A comparison was drawn between specimens processed employing pulsed-Metal Inert Gas and cold metal transfer in terms of hydrogen solubility and metal deposition results.

8.2 EXPERIMENTATION

8.2.1 MATERIALS

The material fabricated by WAAM was developed utilizing ER2618 solid wire and an Al-Mg-Mn alloy wrought plate base, sized at $200 \times 125 \times 20$ mm^3. Table 8.1 lists the materials and their chemical compositions. For this study, Argon gas of 99% purity was employed. The substrate is securely fastened to welding before and during the deposition process to prevent any distortions. Beyond elements presented in Table 8.1, H_2 content in feedstock wire was examined, registering 7.5 ppm per 100g of metal. Prior to such assessments, all wire specimens underwent thorough cleaning and drying procedures. It is noteworthy that hydrogen readings in the wire were influenced by contaminants of the surface. For instance, surface imperfections or unevenness might contribute to organic matter retention.

8.2.2 SAMPLE MANUFACTURING

The impact of deposition parameters on porosity distribution was assessed by crafting eight specimens by traditional pulsed-Metal Inert Gas, and another eight-utilizing cold metal transfer. An OTC Daihen Synchro feed welding system was employed for depositing the pulsed-MIG portion, while the CMT specimens were crafted utilizing a Fronius TPS400i cold metal transfer Advanced power source combined with a Fanuc robot. Based on prior studies (Ghoncheh et al. 2020; Schuster et al. 2023), two heat input levels – maximum and minimum were chosen for techniques with the exact deposition parameters presented in Table 8.2.

Table 8.2 represents mean values derived from gauging the stable mode of metal deposition over roughly 5 s. To ascertain the necessary heat input, Eqs. (1) and (2) were applied (Fathi et al. 2021; Guzmán-Flores et al. 2022). A depiction of fluctuations in parameters such as heat input, voltage, and current can be observed in Figure 8.2, high frequency symbolizes the low frequency and highest heat input denotes the minimum. Each of these specimens consisted of 15 layers with a length of 100 mm each.

$$Heat\ input = \eta \frac{Average\ Voltage \times Average\ Current}{Travel\ Speed} \tag{1}$$

$$Heat\ input = \frac{\eta \sum_{i=1}^{n} \frac{Ii \times Ui}{n}}{Travel\ speed} \tag{2}$$

During the sample production, the interlayer temperatures varied between 60°C and 100°C. Measurements were taken using a K-type digital contact thermometer, chosen based on ASTM E2877 guidelines (Gu et al. 2018). Throughout the component creation, temperature monitoring was solely focused on the topmost layer. After each layer was deposited, temperature measurements were captured at the centre and 25 mm from both extremities along the 100-mm span. The subsequent layer deposition commenced only after the desired temperature, 60°C or 110°C is achieved by

TABLE 8.1

Deposition Wire and Substrate Chemical Composition

Elements	Silicon	Manganese	Chromium	Copper	Titanium	Iron	Zinc	Magnesium	Aluminium
Filler wire Substrate Elements	0.06	0.65	0.07	0.01	0.07	0.14	< 0.01	4.91	Bal
	0.11	0.66	0.06	0.05	0.05	0.25	0.05	4.74	Bal

TABLE 8.2

Manufacturing Parameters Test Specimens

Parameters	Unit	Metal Inert Gas		Cold Metal Transfer	
		Low heat	High heat	Low heat	High heat
Average current	(A)	74	155	74	155
Average voltage	(V)	19.4	19.8	18.3	19.4
Torch travel speed	(m/min)	0.7	0.7	0.7	0.7
Heat input	(J/mm)	161	353	142	347
Wire feed	(m/min)	4.89	9.65	5.1	9.7
Wire feed /travel speed		8.2	15.4	8.2	15.3

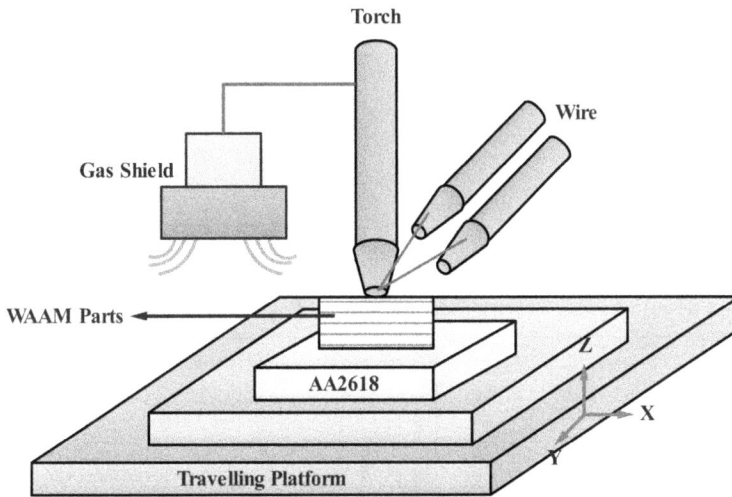

FIGURE 8.1 Schematic of wire arc additive manufacturing.

natural cooling. The base material kept a steady temperature during the initial layer deposition, and the same device was used for continuous temperature monitoring. Eight specimens each were created using both pulsed-MIG and CMT methods. Sets 1 and 2 each consisted of four specimens. A robotic programme was designed with a predetermined interlayer dwell time of 40 s or 130 s, disregarding the interlayer temperature during the deposition of all 15 layers. All the examinations were conducted in a laboratory environment through regulated temperature and humidity to minimize discrepancies between the cold metal transfer and pulsed Metal Inert Gas specimens.

8.2.3 TESTING

Following the creation of the 16 specimens, a segment from each was reduced to approximately 35 mm in length to ensure consistent deposition conditions. The HMX

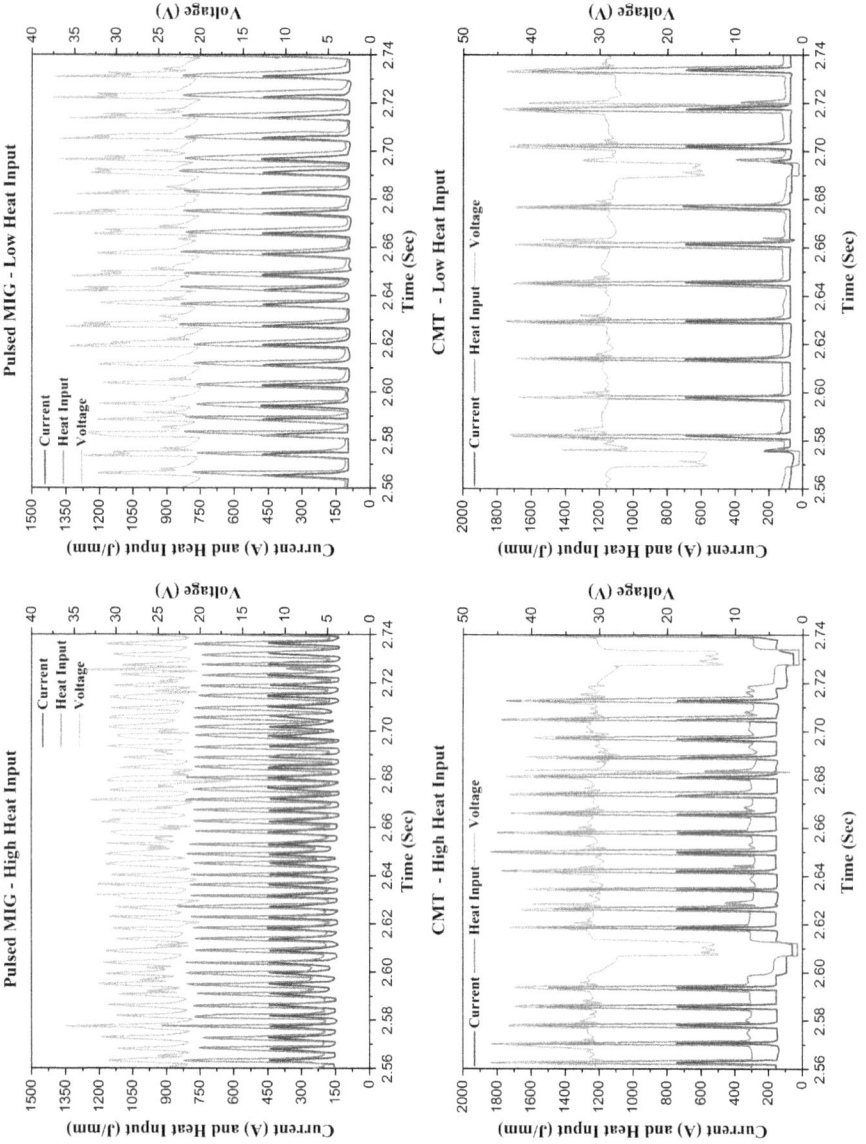

FIGURE 8.2 Pulsed-MIG high-heat-input current, voltage, and heat-input variations.

225 system was utilized to execute X-ray computed tomography (XCT) scans, capturing a volume close to 7200 mm^3 for each scan (Jiangang et al. 2022). Visualization was facilitated through VGStudioMAx software, while operational control and data gathering were managed using X-Tek InspectX.

Post-XCT, segments from the uniform deposition phase were excised for hydrogen dissolution testing (Wahsh et al. 2018; Nagasai, Malarvizhi, & Balasubramanian 2022). The Leco RH402 device ascertained the comprehensive hydrogen content within the specimens. Hydrogen identification tests focused on a characteristic sample from the area that underwent XCT scanning. The evaluations encompassed both dissolved and trapped hydrogen within the specimens. A volume close to 2000 mm^3 of the consistent metal deposition was evaluated for pore comparison and scrutiny.

8.2.4 IDENTIFICATION OF SPECIMENS

In the study, 16 distinct specimens were examined. For clarity, these specimens were assigned specific IDs. For instance, specimen S5 was fabricated using the cold metal transfer method with a high heat input and a mean temperature maintained at 60°C. Conversely, Specimen S11 was generated by the pulsed-Metal Inert Gas approach with a diminished heat input and a pause of 40 s between deposits. Table 8.3 elucidates the organization of specimens crafted under analogous deposition conditions, categorizing them into four separate groups. The fabrication of specimens

TABLE 8.3
Identification of Specimens

Set No.	Metal Deposition Method	Heat Input	Interlayer Temperature	Interlayer Dwell Time	Specimen	Specimen ID
1	Pulsed-Metal Inert Gas (P)	High Heat	60 °C	–	P-HH-60°C	S1
			110 °C	–	P-HH-110°C	S2
		Low Heat	60 °C	–	P-LH-60°C	S3
			110 °C	–	P-LH-110°C	S4
3	Pulsed-Metal Inert Gas (P)	High Heat	–	40 s	P-HH-40s	S9
			–	120 s	P-HH-130s	S10
		Low Heat	–	40 s	P-LH-40 s	S11
			–	120 s	P-LH-130 s	S12
2	Cold Metal Transfer (C)	High Heat	60 °C	–	C-HH-60°C	S5
			110 °C	–	C-HH-110°C	S6
		Low Heat	60 °C	–	C-LH-60°C	S7
			110 °C	–	C-LH-110°C	S8
4	Cold Metal Transfer (C)	High Heat	–	40 s	C-HH-40 s	S13
			–	130 s	C-HH-130s	S14
		Low Heat	–	40 s	C-LH-40 s	S15
			–	130 s	C-LH-130s	S16

within groups 1 and 2 factored in the interlayer temperature, disregarding the pause between each layer. Meanwhile, specimens from groups 3 and 4 were produced with specific intervals between each deposition, while sidelining the consideration of the interlayer temperature.

8.3 RESULTS

8.3.1 VOLUME CONSIDERATION

All 16 specimens underwent identical analysis, resulting in comparable imagery and data on porosity distribution. It is evident that the commencement and termination points of the arc possess a denser concentration of porosity. Both these areas were excluded from the study since they are commonly machined off in the final product. The central focus of this detailed study was a segment representing a constant deposition scenario, located at least 15 mm from both ends and elevated 6 mm from the substrate.

8.3.2 ASSESSMENT OF POROSITY CONTENT

The metal deposition noticeably impacted pore content. Generally, specimens prepared employed by cold metal transfer exhibited reduced pore volumes compared to those produced with pulsed-MIG. The influence of various deposition parameters are heat input, interlayer temperature, and dwell time between layers. CMT, owing to its attributes like dip transfer effects, rapidly oscillating wire, and low heat input inherently produces fewer pores as compared to pulsed-Metal Inert Gas. Specimens generated under conditions of maximum heat input and mean temperature of 100°C displayed a modest 10% disparity in pore between the CMT and pulsed-MIG methods. The most pronounced difference, at 390%, was attained in specimens crafted with a lower heat input and temperature of 60°C. Even for specimens subjected to less heat input and extended dwell time between layers, a stark 360% discrepancy in pore content was evident between the two deposition methods (S16 and S12). Interestingly, the sole pulsed-MIG sample (S9) that exhibited reduced porosity than its CMT counterpart (S13) was generated under conditions of high heat input and a 40-second pause between layers.

Comparing the CMT and pulsed-MIG methods under equivalent processing conditions revealed distinct variances in how each technique reacted to heat input concerning porosity content (Zhao et al. 2022; Gomes et al. 2018). For CMT-produced specimens, there was a uniform rise in porosity, represented as a fraction of the overall volume, when high heat input was applied compared to low heat input. On the other hand, the pulsed-Metal Inert Gas specimens at less heat input exhibited greater pore levels in relation to the total volume than their counterparts that had high heat input. The most pronounced discrepancy in pore volume fraction in the CMT specimens was seen between S5 and S7, both of which had an interlayer temperature set at 60°C. Specimens S4 and S2, with interlayer temperatures fixed at 100 °C, displayed a variation of 93.6% (0.122% for low heat input and 0.063% for high

TABLE 8.4

Calculated Pore Volume as a Function of Either Interlayer Temperature or Interlayer Dwell Time

Specimen ID	Percentage of Pore Volume Fraction based on Specimen Volume
S1	0.106
S2	0.063
S3	0.152
S4	0.122
S5	0.05
S6	0.057
S7	0.031
S8	0.041
S9	0.066
S10	0.127
S11	0.077
S12	0.175
S13	0.07
S14	0.061
S15	0.049
S16	0.038

heat input). Meanwhile, specimens S11 and S9, which had interlayer pauses set at 40 s, registered a 16.6% difference, with values being 0.077% for low heat input and 0.066% for high heat input (Table 8.4).

The study highlighted the significant role that temperature plays in influencing overall porosity. Specifically, in the pulsed-MIG specimens, a lower interlayer temperature correlated with an increase in total porosity across both heat inputs. The disparity in porosity values among high and low heat input specimens was notably substantial at 68.2% and 24.5%, respectively. On the other hand, a contrasting observation was made for CMT specimens. When comparing specimens S6 and S8, which underwent processing at high interlayer temperatures showed increased porosity compared to specimens S5 and S7, which were processed at lower interlayer temperatures.

Furthermore, the duration of interlayer dwell time also displayed a noticeable impact on porosity. Pulsed-MIG specimens produced with a longer interlayer dwell time (120 s) exhibited a greater pore content than their counterparts which had a shorter dwell time (40 s). Specifically, the variation in porosity between the high heat input specimens and the low heat input at 92.4% and 127%, respectively. In the case of CMT specimens, the findings showed a reverse trend. Those with a 40 s dwell time presented a high pore when compared to the 120 s dwell time specimens. The differences here were 14.7% for the high heat input specimens and 28.9% for the low heat input ones.

8.3.3 PORE SIZE

The pore size distribution and percentages with each sample were calculated employing XCT scans. Because their impact on fatigue life was found to be so small, pores smaller than 0.1 mm dia were disregarded. Small pores, medium pores, and large pores were determined. The percentage of total detected pores across all eight specimens is displayed in Table 8.5 by pore size range.

Table 8.5 shows that the small pore population was higher in cold metal transfer than in pulsed-Metal Inert Gas specimens, while the medium and large pore populations were the other way around. Figure 8.3a, b displays this trend, albeit with some modifications, as a purpose of dwell time and temperature. No matter of deposition, small pores made up higher than 60% of the total pore population. However, a sizable number of pores fell somewhere between the two extremes. Pore size was reduced by as much as 77.47% in CMT-made specimens compared to as little as 52.79–63.8% in pulsed-Metal Inert Gas-made specimens. Cold metal transfer specimens had fewer pores in the medium-size range than pulsed-Metal Inert Gas specimens. Specimens processed using cold metal transfer showed only 2035.5 % uniformly sized pores,

TABLE 8.5

Analysis of the Variation in Pore Size between Pulsed Metal Inert Gas and Cold Metal Transfer Aluminium Specimens

Pore dia	Percentage of Pore Count Fraction	
(mm)	Pulsed-Metal Inert Gas	Cold Metal Transfer
Small	53.28 – 64.85	61.78 – 78.65
Medium	33.24 – 43.62	21.2 – 36.45
Large	4.5 – 6.34	2.2 – 5.48

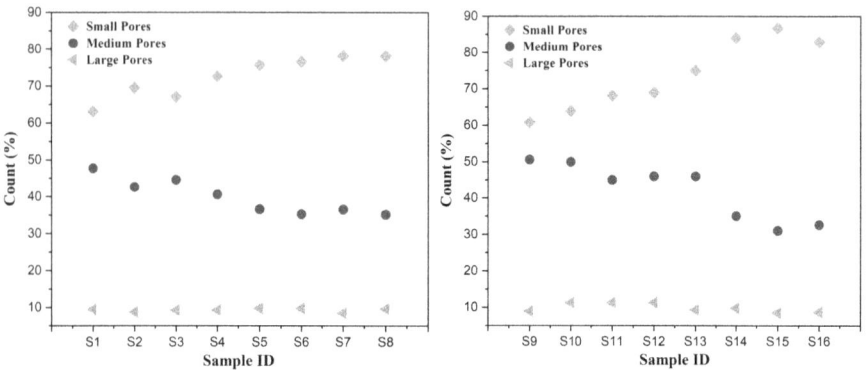

FIGURE 8.3 Number of specimens with varying porosity sizes produced by altering (a) Temperature (b) Dwell time.

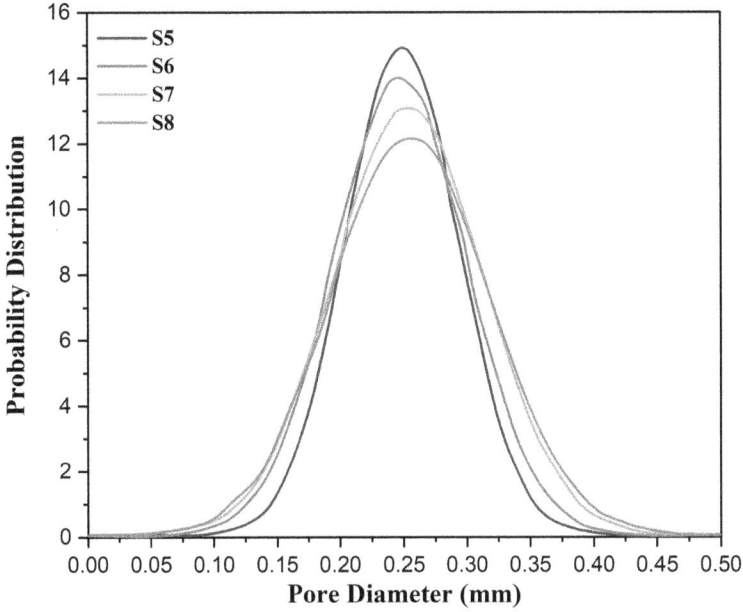

FIGURE 8.4 Normal pore size distribution in cold metal transfer specimens: the impact of interlayer temperature and heat.

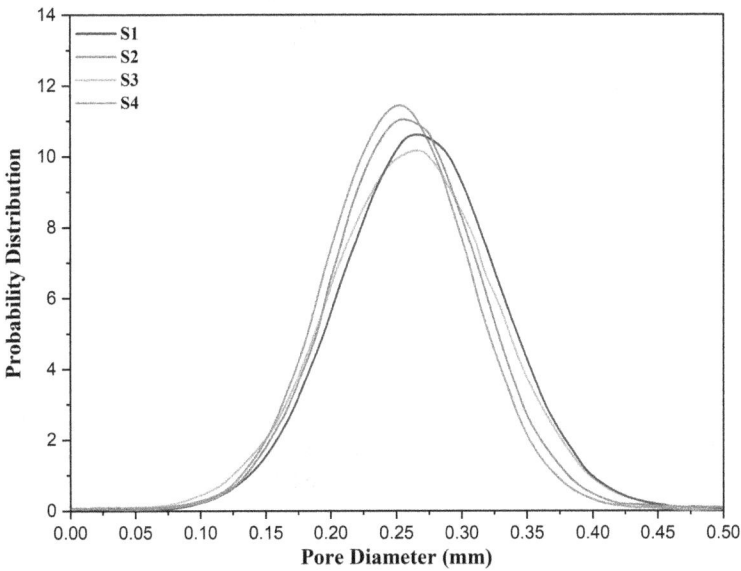

FIGURE 8.5 Pore size normal distribution in pulsed-MIG specimens: the impact of interlayer temperature and heat.

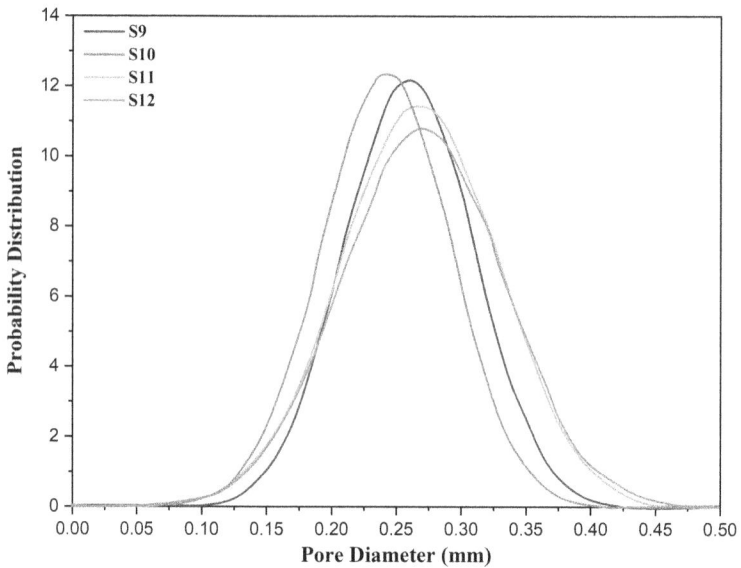

FIGURE 8.6 Influence of metal deposition method on distribution of pore size in high-heat-input specimens produced with varying interlayer dwell times.

TABLE 8.6

Assessment of Total H_2 Content, H_2 in Pores and Dissolved H_2 Attained from Dissolve H_2 Test in Specimens Prepared Using Cold Metal Transfer and Pulsed-Metal Inert Gas

Set ID	Specimen ID	Hydrogen in Specimens of 110 g (ml)	Hydrogen Forming Pores (%)	Hydrogen in Solid Solution (%)
DH1	S7	0.884	1.220	98.674
	S3	1.256	6.20	93.80
DH2	S16	1.220	1.470	98.53
	S12	1.352	5.23	94.77

while those processed using pulsed-MIG showed 32–42%. Pores larger than 0.31 mm in diameter were also more common in the Pulsed-MIG processed specimens re–lated to cold metal transfer specimens (Table 8.6).

8.3.4 DISTRIBUTION OF PORE SIZE

The XCT scan data were analysed to determine the pore size distribution. Pore diameter distribution for low and high heat input CMT specimens, accounted for

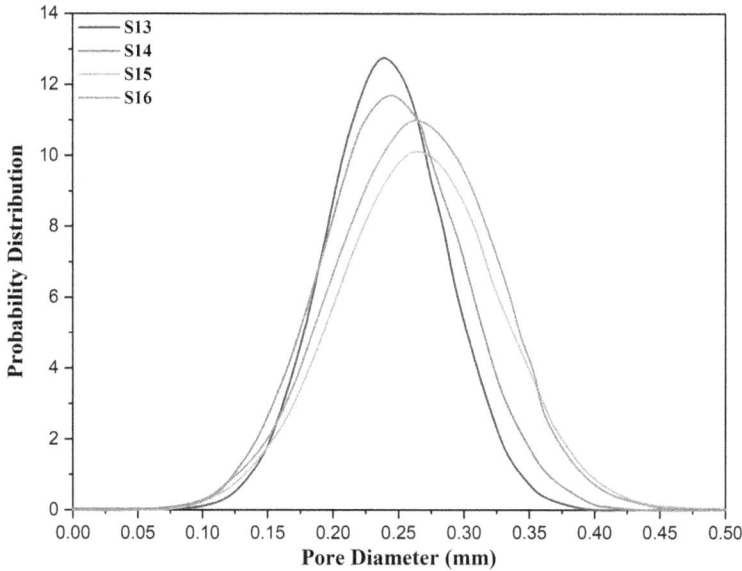

FIGURE 8.7   Pore size distribution in specimens made with low heat input and different interlayer dwell times, and the impact of metal deposition techniques.

interlayer temperatures. The average pore size is roughly the same for all specimens, though it is slightly larger for the high heat input specimens. High-heat-input speci-mens have a more consistent pore size distribution than low-heat-input specimens (Tomar & Shiva 2023; Oyama et al. 2019; Chen, Yang, & Feng 2021). Regardless of the total amount of heat applied, the specimens with a large interpass temperature show the greatest difference in pore size. Similar conditions for pulsed-MIG speci-mens following the effect of the interlayer temperature are in the opposite direction. Lower heat input and lower interlayer temperature are the primary causes of irregu-larities in pulsed-MIG, as evidenced by large pore size and difference in specimens produced with lower interlayer temperature. The gap between pulsed-Metal Inert Gas and cold metal transfer average pore size was small (less than 0.2 mm).

The interlayer dwell time does not significantly affect the pulsed-Metal Inert Gas specimens at either higher or lower heat inputs. Comparing large and less heat input deposition techniques, pulsed-Metal Inert Gas specimens show the greatest varia-tion in pore size, while cold metal transfer processed specimens showed the least. Specimens prepared using cold metal transfer were found to have smaller pores and a distribution of narrower pore size. The average pore size of cold metal transfer specimens was lower than that of pulsed-MIG specimens.

8.3.5   Physical Distribution and Location of Pore

The average pore size, temperature, and dwell time are shown in distribution plots as a function of the deposition technique. In the same conditions, CMT pores were

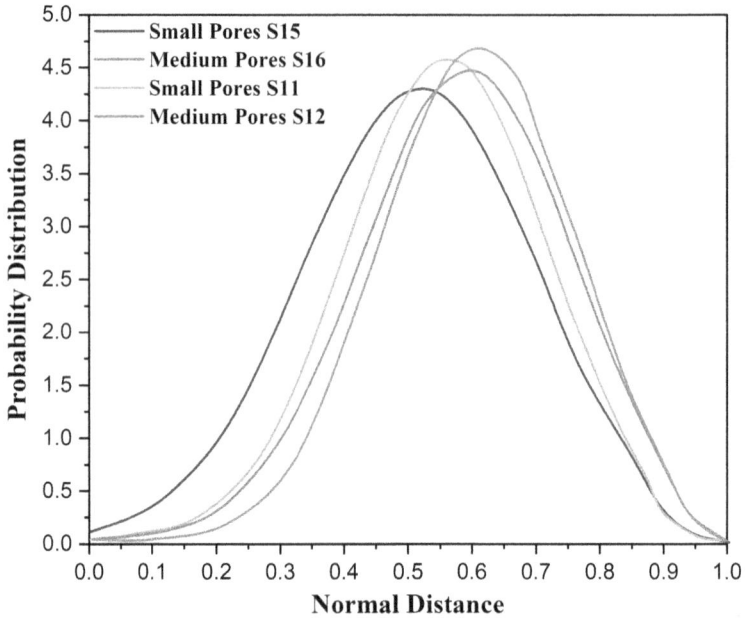

FIGURE 8.8   Normalized pore-to-pore distance as a function of metal deposition technique.

more localized, indicating that they were smaller overall, than pulsed-MIG pores. Furthermore, for CMT, the small pore normal distribution curve was flatter, indicating a greater variation in the normalized pore distance. It was evident that pores of a medium size were dispersed over a wider area than pores of a smaller size. As a result, there was less consistency in the pore count over a relatively narrow region. In the pulsed-MIG sample, pores of all sizes were further away from the sample's centroid on average.

As can be seen in Figure 8.9, the pore size and distribution of CMT specimens are affected by the interlayer temperature. Pores of all sizes were more densely packed in specimens processed at 60°C interlayer temperature compared to those produced at 100°C interlayer temperature. Additionally, there is less variation in the specimens with lower interlayer temperatures, suggesting that the pores were distributed more consistently than in the specimens with higher interlayer temperatures. Similar to Figure 8.8, the normalized distance between the centres of medium-sized pores was larger on average and more variable than that of small pores, suggesting that large pores are distributed less uniformly and at greater distances.

Figure 8.10 illustrates the effect of various heat inputs and reveals that the distribution of pores with medium sizes was more extensive than that of pores with small sizes. Specimens subjected to a high heat input differed more in average normalized distance between groups of small and large pore sizes than specimens subjected to a low heat input. Although there was little to no difference between pore sizes when normalized distances were calculated, smaller pores did exhibit more

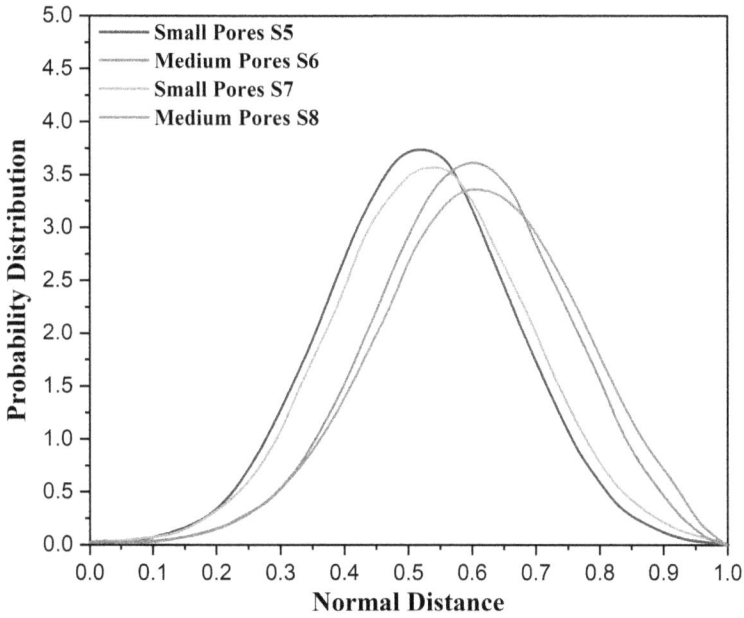

FIGURE 8.9 Impact of temperature on the central tendency of the normalized pore distances away from the mean.

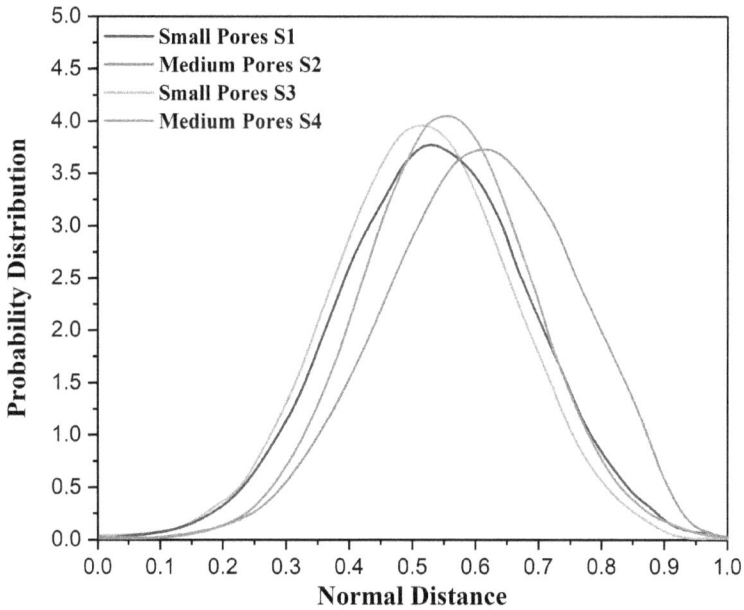

FIGURE 8.10 Normalized pore distances from the pore centroid as a purpose of heat input.

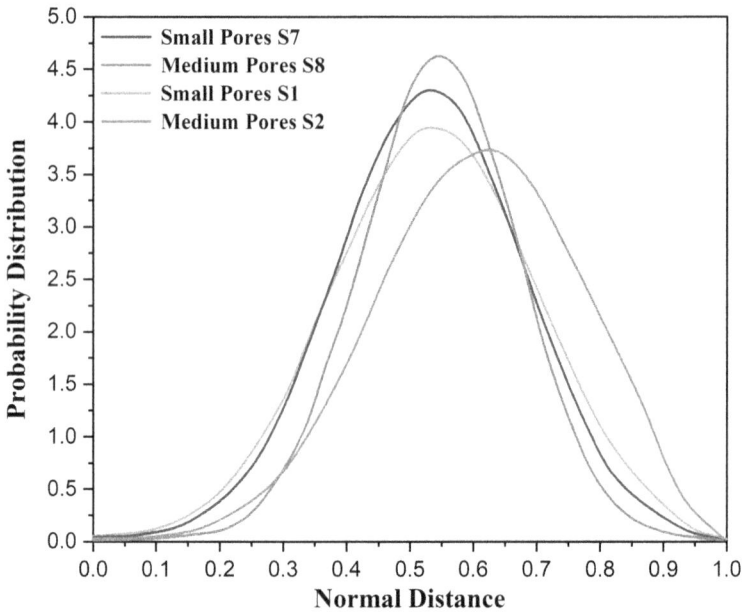

FIGURE 8.11 Normalized pore distances from the pore centroid for two distinct metal deposition conditions.

variation. Interlayer temperatures are shown to be different between the high-heat-input pulsed-MIG sample and the low-heat-input CMT sample as in Figure 8.11.

8.3.6 PORE VOLUME

Small pores accounted for over 60% of pore volume in CMT specimens, which was significantly higher than the percentages of medium and large pore sizes. One sample (S13) stood out from the pack with a total pore volume of 47.3% due to the extreme conditions of high heat input and 40 s interlayer dwell time. Average sized pores accounted for between 31.5% and 44.7% of the total pore volume (Figure 8.12).

However, the results were very different for pulsed-MIG-produced specimens. In every sample except S2 and S4, the total volume of medium-sized pores exceeded the total volume of small-sized pores. Pores with a medium size were found to have a greater total volume in six of the eight specimens compared to pores with a small size. Comparing the pore volumes of S11 and S9, S11 had the smallest percentage difference between medium and small pores (0.91 %) (2.65 %). The answer is yes (9.71 % of the time). Large pore volume in pulsed MIG specimens ranged from 6.8 to 13.1% of the total. Overall, pulsed MIG specimens had a larger percentage of large-pore volume than cold metal transfer specimens. In comparison to CMT specimens, which averaged only 6.1% large-size pores by volume, pulsed-Metal Inert Gas specimens averaged 10.8%.

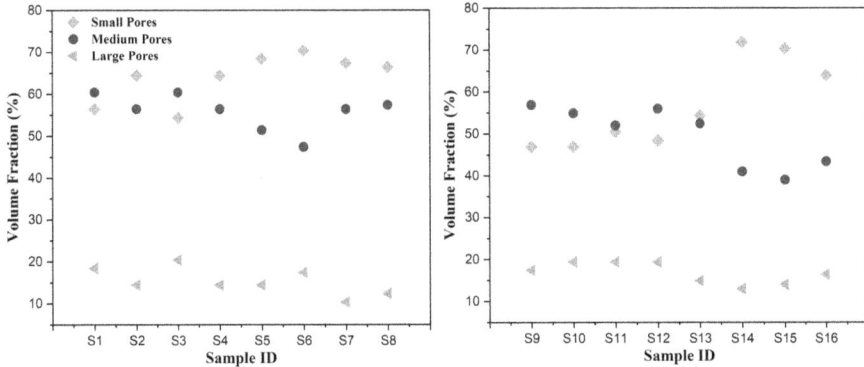

FIGURE 8.12 Variation in porosity size distribution as a function of sample volume produced with varying interlayer temperatures and dwell times.

8.4 DISCUSSION

Temperature control was used during the production of both sample sets 1 and 2, and this process was performed separately from the dwell time. Authors Ding et al. (2016); Tawfik, Nemat-Alla, and Dewidar (2021); and Derekar et al. (2020) explained that as the number of layers increased, the temperature of forming part also increased. The temperature of the deposited layer is lowered by heat extraction of substrate, and this effect is most noticeable for the first few layers. The temperature of the forming component rises as the separation between the deposited layer and the substrate lengthens. Shorter intervals between layer depositions were used to keep the interlayer temperature at a constant value. As a result, in temperature-based specimens, the dwell time is shortest for the first few layers and also gradually raised for the last few layers. The time it takes for liquid metal to cool from 110°C to 60°C is significantly less. It is worth noting at this juncture that specimens prepared with an interlayer temperature of 60°C had a longer interlayer dwell time than those prepared with an interlayer temperature of 100°C. Therefore, it is concluded that specimens made with a temperature of 100°C are consistently more heated/hotter than specimens made with an interlayer temperature of 60°C, which provided more time for heat to be dissipated into the environment.

Alternatively, dwell times of either 40s or 130 s were used during sample production regardless of the top layer temperature. Heat accumulated more quickly when the interlayer dwell time was only 40 s, compared to when it was 130 s, because less heat could be dissipated to the substrate and surroundings during the shorter interlayer dwell time. Therefore, the dwell time of 40 s resulted in a higher interlayer temperature than the dwell time of 130 s. Specimens prepared with an interlayer temperature of 60°C were inferred to be roughly equivalent to those prepared with a dwell time of 130 s, all else being equal. What's more, it's possible that specimens created with a 40-second interlayer dwell time are on par with those created with an interlayer temperature of 110°C.

8.4.1   PENETRATION EFFECT AND DEPOSITION METHOD

In CMT, the arc is turned on and off repeatedly while the feed wire moves forward and backward under electronic control to lower the arc's energy and input of heat. The amount of melting and penetration is reduced as a result of this. Pulsed-MIG welding maintains an arc at all times without experiencing wire retraction, despite current pulsing reducing the total arc energy. The penetration diagram suggests that pores formed on top of the deposited layer are sealed off entirely by the arc penetration effect during the deposition of the next layer. When a layer is melted away, any pores that happen to be in that area are also destroyed. When a new layer is deposited on top of an existing one, its pores will fill in because they are located within the layer's penetration zone. However, pores formed at the interlayer region and the base of a layer are unaffected. This is because there was insufficient depth penetration to reach the full depth of the layer. Both CMT and pulsed-MIG specimens exhibit pronounced pore banding along the length of the layer in the region. Pulsed-MIG welding offers more opportunities for hydrogen absorption than CMT welding because the arc and liquid aluminium are operated at higher temperatures.

According to author Fang et al. (2018), solute banding in welds is similar to the "banding zones" that form in pores during MIG welding of aluminium. It is concluded that a significant factor in banding formation is periodic variation in solidification rate. Increased pores at interlayer regions may also result from solid state porosity formation, also known as secondary porosity, which can occur as a result of succeeding reheating near solid temperature.

8.4.2   ABSORBED HYDROGEN

Only when the H_2 concentration in the liquid exceeds the high soluble limit of hydrogen in solid AA2618 do hydrogen bubbles form. Maximum hydrogen solubility is reached at the centre of the arc during MIG welding, and the hydrogen is then diffused throughout the weld pool by convection. The percentage of hydrogen available for pore formation is significantly higher in cold metal transfer processed specimens compared to pulsed-Metal Inert Gas specimens.

H2 in the pores of pulsed-Metal Inert Gas specimens was higher than CMT specimens in both sets. Dissolved hydrogen levels in pulsed-Metal Inert Gas specimens were around 95%, while cold metal transfer levels were above 98%. This is in addition to the hydrogen that is trapped in the pores. Total hydrogen measured during testing was assumed to exist as dissolved atoms in the lattice imperfections for all these calculations. Because of this, a smaller percentage of hydrogen was left in a dissolved state after pore formation increased in pulsed MIG. Thus, despite forming fewer pores and absorbing less atmospheric hydrogen than pulsed-Metal Inert Gas specimens, cold metal transfer demonstrated higher dissolved H_2 for a variety of reasons.

While arc length remains relatively constant during the metal deposition process when using pulsed-MIG, it fluctuates between maximum and zero when using CMT due to the short-circuiting mode. The aluminium liquid can then absorb more hydrogen from the surrounding air or a protective gas. The droplets formed and transferred during CMT likely had a smaller surface area than those formed and transferred during pulsed-MIG spraying.

8.4.3 Impact of Solidification and Cooling Rate

Pore sites and lattice imperfections like grain boundaries, dislocations, impurities, etc. are thought to be sources of hydrogen in aluminium. The rejection of dissolved hydrogen causes pores to form at the solidification front during the solidification of liquid metal. Dissolved gas must be allowed to escape from liquid metal before it is rejected. The probability that dissolved hydrogen will be preserved within the solid metal increases as the solidification rate rises. CMT, with its unconventional approach to metal deposition, allows metal to solidify at a quicker rate than pulsed-MIG for the processes used in this investigation. This led researchers to conclude that aluminium specimens produced using CMT had a higher probability of retaining dissolved hydrogen than those produced using pulsed-MIG (Chen et al. 2022; Shen, Kong, & Chen 2021). This is consistent with the idea that hydrogen ions dissolved in the molten aluminium give increment to the formation of pores during solidification, pulsed-Metal Inert Gas specimens had a higher total pore volume than CMT specimens. Due to the slow solidification rate during pulsed-MIG processing, hydrogen is more likely to be picked up by the molten metal. Pulsed-MIG specimens might have a higher total hydrogen content because of this.

Hydrogen bubbles can form in the spaces between developing dendrites, but their ability to detach and float into the nearby liquid aluminium is contingent on the size and shape of these interstices. Hydrogen bubbles between forming cells cannot detach and grow at a faster cooling rate because their growth rate is slower than that of the progressing closely packed cells. The large gap between dendrites allows for a larger number of pores to fill the available space, reducing the rate at which the cell cools. This is a possible contributing factor to the larger pore sizes observed in pulsed-Metal Inert Gas specimens compared to cold metal transfer specimens.

When metal is deposited in layers, the temperature of the deposit rises with each successive layer. An area's temperature rise depends on its proximity to the deposit's top layer, the thermal conductivity of the alloy composition, and the arc energy. Authors Y. Zhang et al. (2022); J. Zhang, Xing, and Cao (2022); and Mclean et al. (2022) discussed the temperature distribution that occurs during metal deposition. When depositing a metal onto another layer, the temperature of the substrate must be raised above the recrystallization temperature of the metal.

Aluminium concentration and dissolved hydrogen concentration influence the diffusion of hydrogen, as shown by authors Jia et al. (2023) and Klein and Schnall (2020) at high and low temperatures. Hydrogen diffusion may have been marginally enhanced in the pulsed-Metal Inert Gas specimens compared to the cold metal transfer specimens as a result of the higher arc energy, deeper penetration, and forced vacancy diffusion. CMT's faster solidification, shallower penetration, and lower arc energy may have less of an impact on hydrogen diffusion than pulsed-MIG. The larger pores observed in pulsed-MIG specimens may have been caused by the formation of clusters, which in turn may have been caused by the freer movement of hydrogen and vacancies. This finding clarified why pulsed MIG specimens had larger pore sizes and a higher volume fraction of large pore sizes than cold metal transfer specimens.

The total pore volume of pulsed-Metal Inert Gas specimens was found to be greater for lower heat input and lower interlayer temperature control methods. Therefore, pore volume in specimens was diminished under conditions of higher heat input and higher temperature. The total pore volume of the CMT specimens was high when the heat input and interlayer temperature were both raised, and low when the heat input and interlayer temperature were lowered. These apparently contradictory results suggest a link between the deposit's heat content and the solidification and coalescence processes, which in turn produce pores. This research shows that the high heat input and interlayer temperature present during pulsed-Metal Inert Gas deposition mode can support hydrogen coalescence, resulting in the formation of pores through which hydrogen can escape. Due to the low heat input and interlayer temperature in CMT, hydrogen cannot coalesce and escape the pores even at higher temperatures. The conditions for pore formation in both processes provide enough heat for H_2 formation, but not enough heat to release hydrogen as pores. CMT is much easier on the hands than pulsed-MIG, so keep that in mind.

8.4.4 STATISTICAL ANALYSIS

To confirm the variances in porosity diameteroccurred during fabrication of the specimens under varying metal deposition conditions, a statistical analysis called ANOVA was performed. The significance levels from the analysis of variance were taken into account for the analysis.

It is a common null hypothesis to assume that specimens have the same porosity diameter. The null hypothesis is rejected if the p value for the difference in porosity diameters between the specimens is less than 0.05. The porosity diameters of specimens created with a pulsed-Metal Inert Gas process are not significantly different from one another, as shown in Table 8.7 by the lack of any p values below 0.05. In contrast, the porosity diameter is highly sensitive to the varying parameters used in a CMT process. Table 8.7 shows the CMT conditions where heat input produces statistically significant differences in diameters, while specimens with varying interlayer temperature and interlayer dwell time display only small differences. When the heat inputs are small, the probability that the null hypothesis is incorrect due to different interlayer temperatures is 86.13%, and when the heat inputs are large, the probability drops to 64.09%.

Table 8.9 shows that when comparing low heat input specimens from cold metal transfer and pulsed-Metal Inert Gas processes, the null hypothesis was rejected with a probability significantly lower than 0.05. This indicates that low heat input conditions have a greater effect on pore diameter in cold metal transfer and pulsed process specimens than high heat input conditions. Although statistical results suggested a correlation between CMT and pulsed-MIG pore production under high heat input conditions, confidence validity of this hypothesis was low.

TABLE 8.7

Comparing Cold Metal Transfer and Pulsed-Metal Inert Gas Deposition

	CMT		P-MIG	
Comparison	Specimen Id	p Values	Specimen Id	p Values
Interlayer temperature	S5	0.3602	S1	0.563
	S6		S2	
	S7	0.1391	S3	0.7623
	S8		S4	
Interlayer dwell time	S13	0.362	S9	0.634
	S14		S10	
	S15	0.2253	S11	0.6326
	S16		S12	
Heat input	S5	1.2×10^{-38}	S1	0.2674
	S7		S3	
	S6	4.51×10^{-40}	S2	0.3874
	S8		S4	
	S13	1.43×10^{-75}	S9	0.6672
	S15		S11	
	S14	2.95×10^{-44}	S10	0.4663
	S16		S12	

TABLE 8.8

The p-Values for Metal Deposition Parameters Compared between the CMT and Pulsed-MIG

Conditions		Sample ID	p Values
High heat input	Temperature	S5	0.3222
		S1	0.251
		S6	
		S2	
	Dwell time	S13	0.3884
		S9	0.1177
		S14	
		S10	
Low heat input	Temperature	S7	1.25×10^{-37}
		S3	3.71×10^{-38}
		S8	
		S4	
	Dwell time	S15	4.63×10^{-91}
		S11	5.24×10^{-86}
		S16	
		S12	

TABLE 8.9

Statistical Significance of Metal Deposition Parameters Obtained from Various Sample Configurations

Condition	Specimen IDs	p Values
Max. heat content	S2	1.52×10^{-40}
	S7	
Equivalent heat content	S3	0.0341
	S6	
Equivalent condition of	S5	0.1033
temperature and time based	S14	0.094
specimens of CMT and P-MIG	S6	
	S13	1.6×10^{-12}
	S7	
	S16	6.29×10^{-29}
	S8	
	S15	0.2722
	S1	
	S10	0.6486
	S2	
	S9	0.711
	S3	
	S12	0.2712
	S4	
	S11	

Similar temperatures were used in a second scenario where it was discovered that CMT and pulsed-MIG both produced pores of varying sizes (specimens S3 and S6) (Table 8.8). The comparison of specimens that were interleaved at 60°C and 110°C with those that were interleaved at 130 s and 40 s. Mathematical invariance was observed for CMT low heat input conditions within the same class; however, a pattern did emerge.

8.5   CONCLUSIONS

1. The pore content of pulsed-Metal Inert Gas was consistently larger than that of CMT. In both cases, the vast majority of pores were quite tiny, with diameters between 0.11 mm and 0.20 mm. Pulsed-MIG, in contrast to CMT, had a greater proportion of pores between 0.21 and 0.40 microns in diameter and larger than 0.31 microns in size. Pulsed-MIG had a greater pore volume for medium-sized pores, while CMT had a greater pore volume for small-sized pores.
2. Higher arc energy meant a bigger, hotter, and slower cooling melt pool, making pulsed-MIG more susceptible to hydrogen absorption than CMT.

3. When compared to CMT, pulsed-MIG was able to retain a smaller fraction of H_2 in solid solution. During the solidification phase, the remaining H_2 was used up in the process of pore formation.
4. Compared to high interlayer temperature, heat input, and low dwell time, low heat input, less temperature, and high dwell time produced a pore volume fraction for pulsed-Metal Inert Gas. It was the complete opposite for CMT.

REFERENCES

Aldalur, E., Suárez, A., & Veiga, F. (2021). Metal transfer modes for wire arc additive manufacturing Al-Mg alloys: Influence of heat input in microstructure and porosity. *Journal of Materials Processing Technology*, *297*, 117271.

Chen, D., Wang, L., He, S., Lyu, F., & Zhan, X. (2022). Three-dimensional forming characteristics and mechanical property of additive manufacturing aluminium–copper alloys. *Materials Science and Technology*, *38*(17), 1519–1531.

Chen, F., Yang, Y., & Feng, H. (2021). Regional control and optimization of heat input during CMT by wire arc additive manufacturing: Modeling and microstructure effects. *Materials*, *14*(5), 1061.

Derekar, K. S., Griffiths, D., Joshi, S. S., Lawrence, J., Zhang, X., Addison, A., & Xu, L. (2020). Influence of interlayer temperature on microstructure of 5183 aluminium alloy made by wire arc additive manufacturing. *International Journal of Microstructure and Materials Properties*, *15*(4), 267–286.

Ding, D., Pan, Z., Van Duin, S., Li, H., & Shen, C. (2016). Fabricating superior NiAl bronze components through wire arc additive manufacturing. *Materials*, *9*(8), 652.

Fang, X., Zhang, L., Chen, G., Dang, X., Huang, K., Wang, L., & Lu, B. (2018). Correlations between microstructure characteristics and mechanical properties in 5183 aluminium alloy fabricated by wire-arc additive manufacturing with different arc modes. *Materials*, *11*(11), 2075.

Fathi, P., Rafieazad, M., Mohseni-Sohi, E., Sanjari, M., Pirgazi, H., Amirkhiz, B. S., & Mohammadi, M. (2021). Corrosion performance of additively manufactured bimetallic aluminum alloys. *Electrochimica Acta*, *389*, 138689.

Gao, Z., Li, Y., Shi, H., Lyu, F., Li, X., Wang, L., & Zhan, X. (2023). Microstructure characteristics under varying solidification parameters in different zones during CMT arc additive manufacturing process of 2319 aluminum alloy. *Vacuum*, *214*, 112177.

Ghoncheh, M. H., Sanjari, M., Cyr, E., Kelly, J., Pirgazi, H., Shakerin, S., … Mohammadi, M. (2020). On the solidification characteristics, deformation, and functionally graded interfaces in additively manufactured hybrid aluminum alloys. *International Journal of Plasticity*, *133*, 102840.

Gomes, B. F., Morais, P. J., Ferreira, V., Pinto, M., & de Almeida, L. H. (2018). Wire-arc additive manufacturing of Al-Mg alloy using CMT and PMC technologies. In *MATEC Web of Conferences* (Vol. 233, p. 00031). EDP Sciences.

Gu, J., Bai, J., Ding, J., Williams, S., Wang, L., & Liu, K. (2018). Design and cracking susceptibility of additively manufactured Al-Cu-Mg alloys with tandem wires and pulsed arc. *Journal of Materials Processing Technology*, *262*, 210–220.

Guzmán-Flores, I., Granda-Gutiérrez, E. E., Martínez-Delgado, D. I., Acevedo-Dávila, J. L., Díaz-Guillén, J. C., Vargas-Arista, B., & Cruz-González, C. E. (2022). Mechanical performance and failure mechanism of layered walls produced by wire arc additive manufacturing in metal transfer pulsed mode. *Journal of Materials Engineering and Performance*, *31*(10), 8522–8530.

Han, X., Yang, Z., Ma, Y., Shi, C., & Xin, Z. (2020). Comparative study of laser-arc hybrid welding for AA6082-T6 aluminum alloy with two different arc modes. *Metals*, *10*(3), 407.

Huang, Z., Huang, J., Yu, X., Liu, G., & Fan, D. (2023). The microstructure and mechanical properties of high-strength 2319 aluminum alloys fabricated by wire double-pulsed metal inert gas arc additive manufacturing. *Journal of Materials Engineering and Performance*, *32*(4), 1810–1823.

Hussein, N. I. S., Ket, G. C., Rahim, T. A., Ayof, M. N., Abidin, M. Z. Z., & Srithorn, J. (2023). Process and heat resources for wire arc additive manufacturing of aluminium alloy ER4043: A review. *Journal of Mechanical Engineering (1823-5514)*, *20*(1), 21–41.

Jia, C., Song, Y., Wang, W., Wang, Y., Wei, Z., & Sun, Z. (2023). A survey of wire arc additive manufacturing technologies for metal materials. *Xiyou Jinshu/Chinese Journal of Rare Metals*, *47*(5), 633–646.

Jiangang, P., Bo, Y., Jinguo, G., Liang, Z., & Hao, L. (2022). Influence of arc mode on the microstructure and mechanical properties of 5356 aluminum alloy fabricated by wire arc additive manufacturing. *Journal of Materials Research and Technology*, *20*, 1893–1907.

Kalman, H., Tripathi, N. M., Gabrieli, O. G., & Portnikov, D. (2017). Phase diagrams for pneumatic and hydraulic conveying. In *18th International Conferences on Transport and Sedimentation of Solid Particles, T and S 2017* (pp. 145–152). Wydawnictwo Uniwersytetu Przyrodniczego we Wroclawiu.

Klein, T., & Schnall, M. (2020). Control of macro-/microstructure and mechanical properties of a wire-arc additive manufactured aluminum alloy. *The International Journal of Advanced Manufacturing Technology*, *108*, 235–244.

Lumay, G., Tripathi, N. M., & Francqui, F. (2019). How to gain a full understanding of powder flow properties, and the benefits of doing so. *ONdrugDelivery*, *2019*, 42–47.

Manikandan, R., Ponnusamy, P., Nanthakumar, S., Gowrishankar, A., Balambica, V., Girimurugan, R., & Mayakannan, S. (2023). Optimization and experimental investigation on AA6082/WC metal matrix composites by abrasive flow machining process. *Materials Today: Proceedings*.

Mayakannan, S., Rathinam, R., Saminathan, R., Deepalakshmi, R., Gopal, M., Hillary, J., & Singh, P. (2022). Analysis of spectroscopic, morphological characterization and interaction of dye molecules for the surface modification of TiB 2 nanoparticles. *Journal of Nanomaterials*, *2022*. https://doi.org/10.1155/2022/1033216

Mclean, N., Bermingham, M. J., Colegrove, P., Sales, A., & Dargusch, M. S. (2022). Understanding the grain refinement mechanisms in aluminium 2319 alloy produced by wire arc additive manufacturing. *Science and Technology of Welding and Joining*, *27*(6), 479–489.

M Tripathi, N., & S Mallick, S. (2017). Pneumatic conveying of Fly Ash: Bend Models investigation. *Advanced Materials Proceedings*, *2*(8), 526–531.

Nagasai, B. P., Malarvizhi, S., & Balasubramanian, V. (2022). Mechanical properties and microstructural characteristics of AA5356 aluminum alloy cylindrical components made by wire arc additive manufacturing process. *Materials Performance and Characterization*, *11*(1), 73–98.

Neveu, A., Lumay, G., Pillitteri, S., Monsuur, F., Pauly, T., Ribeyre, Q., Francqui, F., Vandewalle, N., & Tripathi, N. M. (2020b). Physical characterization of blends containing mesoporous particles with a focus on electrostatic properties. In *2020 AIChE Spring Meeting and 16th Global Congress on Process Safety*. AIChE.

Neveu, A., Tripathi, N. M., Rigo, O., Francqui, F., & Lumay, G. (2020a). Experimental investigation of spreadability of metal powders in recoating process. In *Proceedings - Euro PM2020 Congress and Exhibition (Proceedings - Euro PM2020 Congress and Exhibition)*. European Powder Metallurgy Association (EPMA).

Oyama, K., Diplas, S., M'hamdi, M., Gunnæs, A. E., & Azar, A. S. (2019). Heat source management in wire-arc additive manufacturing process for Al-Mg and Al-Si alloys. *Additive Manufacturing*, *26*, 180–192.

Ponomareva, T., Ponomarev, M., Kisarev, A., & Ivanov, M. (2021). Wire arc additive manufacturing of Al-Mg alloy with the addition of scandium and zirconium. *Materials*, *14*(13), 3665.

Ratsimba, A., Zerrouki, A., Tessier-Doyen, N., Nait-Ali, B., André, D., Duport, P., Neveu, A., Tripathi, N., Francqui, F., & Delaizir, G. (2021). Densification behaviour and three-dimensional printing of Y2O3 ceramic powder by selective laser sintering. *Ceramics International*, *47*(6), 7465–7474. https://doi.org/10.1016/j.ceramint.2020.11.087

Satishkumar, P., Krishnan, G. G., Seenivasan, S., & Rajarathnam, P. (2021). A study on tribological evaluation of hybrid aluminium metal matrix for thermal application. *Materials Today: Proceedings*. https://doi.org/10.1016/j.matpr.2021.04.389

Schuster, M., De Luca, A., Kucajda, D., Hosseini, E., Widmer, R., Maeder, X., & Leinenbach, C. (2023). Heat treatment response and mechanical properties of a Zr-modified AA2618 aluminum alloy fabricated by laser powder bed fusion. *Journal of Alloys and Compounds*, *962*, 171166.

Sharma, A., Babbar, A., Tian, Y., Pathri, B.P., Gupta, M., & Singh, R. (2022). Machining of ceramic materials: A state of the art review. *International Journal on Interactive Design and Manufacturing (IJIDeM)*, 1–21. https://doi.org/10.1007/s12008-022-01016-7

Sharma, A., & Jain, V. (2020). Experimental investigation of cutting temperature during drilling of float glass specimen. In *IOP Conference Series: Materials Science and Engineering* (Vol. 715, No. 1, p. 012050). IOP Publishing.

Sharma, A., Jain, V., & Gupta, D. (2018). Characterization of chipping and tool wear during drilling of float glass using rotary ultrasonic machining. *Measurement*, *128*, 254–263.

Sharma, A., Jain, V., & Gupta, D. (2019a). Tool wear analysis while creating blind holes on float glass using conventional drilling: A multi-shaped tools study. In *Advances in Manufacturing Processes: Select Proceedings of ICEMMM 2018* (pp. 175–183). Springer Singapore.

Sharma, A., Jain, V., & Gupta, D. (2019b). Comparative analysis of chipping mechanics of float glass during rotary ultrasonic drilling and conventional drilling: For multi-shaped tools. *Machining Science and Technology*, *23*(4), 547–568.

Sharma, A., Jain, V., & Gupta, D. (2019c). Multi-shaped tool wear study during rotary ultrasonic drilling and conventional drilling for amorphous solid. *Proceedings of the Institution of Mechanical Engineers, Part E: Journal of Process Mechanical Engineering*, *233*(3), 551–560.

Sharma, A., Jain, V., & Gupta, D. (2021a). Effect of pre and post tempering on hole quality of float glass specimen: For rotary ultrasonic and conventional drilling. *Silicon*, *13*, 2029–2039.

Sharma, A., Jain, V., & Gupta, D. (2021b). Mathematical approach on chipping volume estimation generated during rotary ultrasonic drilling for float glass. *Proceedings of the National Academy of Sciences, India Section A: Physical Sciences*, *92*, 285–291.

Sharma, A., Jain, V., Gupta, D., & Babbar, A. (2020). A review study on miniaturization. In *Advanced Manufacturing and Processing Technology* (First edition, pp. 111–131). CRC Press.

Shen, Q., Kong, X., & Chen, X. (2021). Fabrication of bulk Al-Co-Cr-Fe-Ni high-entropy alloy using combined cable wire arc additive manufacturing (CCW-AAM): Microstructure and mechanical properties. *Journal of Materials Science & Technology*, *74*, 136–142.

Srinivasan, R., Karunakaran, S., Hariprabhu, M., Arunbharathi, R., Suresh, S., Nanthakumar, S.,& Jayakumar, M. (2023). Investigation on the Mechanical Properties of Powder Metallurgy-Manufactured AA7178/ZrSiO 4 Nanocomposites. *Advances in Materials Science and Engineering*, *2023*, 1–11.

Tawfik, M. M., Nemat-Alla, M. M., & Dewidar, M. M. (2021). Effect of travel speed on the properties of Al-Mg aluminum alloy fabricated by wire arc additive manufacturing. *Journal of Materials Engineering and Performance*, *30*(10), 7762–7769.

Tomar, B., & Shiva, S. (2023). Cold metal transfer-based wire arc additive manufacturing. *Journal of the Brazilian Society of Mechanical Sciences and Engineering*, *45*(3), 157.

Tripathi, N. M., Francqui, F., & Lumay, G. (2020). Influence of relative air humidity on the flow property of fine powders. In *Third International Conference on Powder, Granule and Bulk Solids: Innovations and Applications PGBSIA 2020 February 26–28, 2020* (p. 63).

Tripathi, N. M., Francqui, F., Pirenne, T., & Lumay, G. (2019). Measuring food powders electrical properties as a result of anti-static content. In *2019 AIChE Annual Meeting*. American Institute of Chemical Engineers.

Tripathi, N. M., Levy, A., & Kalman, H. (2016), Initial acceleration pressure drop in dilute phase pneumatic conveying system. In *Powder, Granule and Bulk Solids: Innovations and Applications Conference*.

Tripathi, N. M., & Mallick, S. S. (2014). *An Investigation into Pressure Drop Across Bends for Fluidised Densephase Pneumatic Conveying Systems* (Doctoral dissertation).

Wahsh, L. M., ElShater, A. E., Mansour, A. K., Hamdy, F. A., Turky, M. A., Azzam, M. O., & Salem, H. G. (2018). Parameter selection for Wire Arc Additive Manufacturing (WAAM) process. *Materials Science and Technology*, *1*, 78–85.

Wieczorowski, M., Pereira, A., Carou, D., Gapinski, B., & Ramírez, I. (2023). Characterization of 5356 aluminum walls produced by Wire Arc Additive Manufacturing (WAAM). *Materials*, *16*(7), 2570.

Xian, G., Oh, J. M., Lee, J., Cho, S. M., Yeom, J. T., Choi, Y., & Kang, N. (2022). Effect of heat input on microstructure and mechanical property of wire-arc additive manufactured Ti-6Al-4V alloy. *Welding in the World*, *66*(5), 847–861.

Zhang, J., Xing, Y., & Cao, J. (2022). Effect of ultrasonic vibration on microstructure and properties of aluminum alloy produced by CMT wire arc additive manufacturing. *Jinshu Rechuli/Heat Treatment of Metals*, *47*(4), 159–164.

Zhang, Y., Gao, M., Lu, Y., & Du, W. (2022). Deposition geometrical characteristics of wire arc additive-manufactured AA2219 aluminium alloy with cold metal transfer pulse advance arc mode. *The International Journal of Advanced Manufacturing Technology*, *123*(11–12), 3807–3818.

Zhao, W. Y., Cao, X. Y., Du, X. W., Wei, Y., Liu, R., & Chen, J. (2022). Numerical simulation of heat and mass transfer in CMT-based additive manufacturing. *Journal of Mechanical Engineering*, *58*(01), 267–276.

9 A Comprehensive Review on the Machining Process

Unconventional as an Alternative to Conventional Machining

Rahutosh Ranjan, Dharminder Singh,
Vikas Pandey, Ankit Sharma,
A.S.K. Sinha, and Naveen Mani Tripathi

9.1 INTRODUCTION TO MANUFACTURING

Recently, the ever-increasing demand for new-generation materials and alloys having greater mechanical and physical properties has accelerated the advancement in the manufacturing industry. Manufacturing is the process of modification of materials and alloys into functional products. The manufacturing process does not alter the physical and mechanical properties of the materials during the modification of materials shape and size. Thus, the main aim of the manufacturing industry is to provide a high-quality product quickly and cheaply. In addition, the miniaturization of devices and interactive research between scientists from different domains have driven the innovation and adoption of advanced technology in the manufacturing industry. Basically, manufacturing is a manifestation of several steps such as designing, material selection, planning, inventory control, quality assurances, and marketing. Further, manufacturing can also be defined through both technological as well as economic contexts. The former alters the geometry (shape and size) of the workpiece and the latter adds value to the products manufactured, as shown in Figure 9.1.

9.2 MANUFACTURING PROCESS

The manufacturing process plays a vital role in the production of useful products that we use in our daily life. Manufacturing process is a unit operation that increases

DOI: 10.1201/9781003428862-9

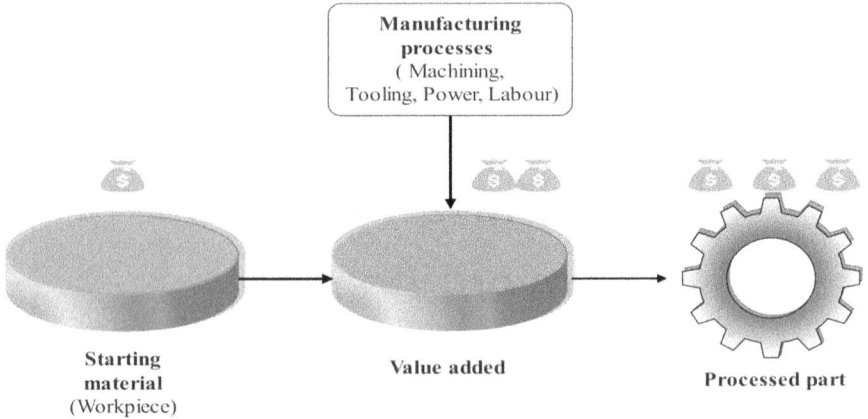

FIGURE 9.1   Technological as well as economic definition of manufacturing.

the utilities of materials in different sectors of life. The manufacturing process is executed in several operations as shown in Figure 9.2. Several manufacturing processes thereby collectively form manufacturing systems and thus produce products from the inputs of the consumers. Both discrete (individual parts) products as well as continuous products can be produced during the manufacturing process. Individual parts such as gears, nails, metal balls, bottles, etc. can be produced and these parts can be used as individual parts as well as assembled to form new devices. Similarly, the production of continuous products such as plastic or metal sheets, pipes, and wires are produced. Therefore, the final product undergoes numerous changing processes, which add value to the final item. The additive and subtractive manufacturing processes show the effects of the parametric conditions (Sharma et al., 2022; Sharma & Jain, 2020; Sharma et al., 2018; Sharma et al., 2019a; Sharma et al., 2019b; Sharma et al., 2019c; Sharma et al., 2021a; Sharma et al., 2021b; Sharma et al., 2020; Sharma et al., 2022; Kalman et al., 2017; Lumay et al., 2019; Tripathi & Mallick, 2017; Neveu et al., 2020a; Neveu et al., 2020b; Ratsimba et al., 2021; Tripathi et al., 2020; Tripathi et al., 2019; Tripathi et al., 2016; Tripathi & Mallick, 2014).

9.2.1   CLASSIFICATION OF MANUFACTURING PROCESSES

The manufacturing process is classified into three sub-operations:
   1) **Processing Operations**: The processing operation changes the shape or physical properties of materials by using mechanical, thermal, electrical, and chemical energy. Generally, discrete products are being produced during processing operations, which can be assembled in further manufacturing processes. However, processing operation results in the generation of waste or chips and the deformation of shapes during material removal and casting processes, respectively. The processing

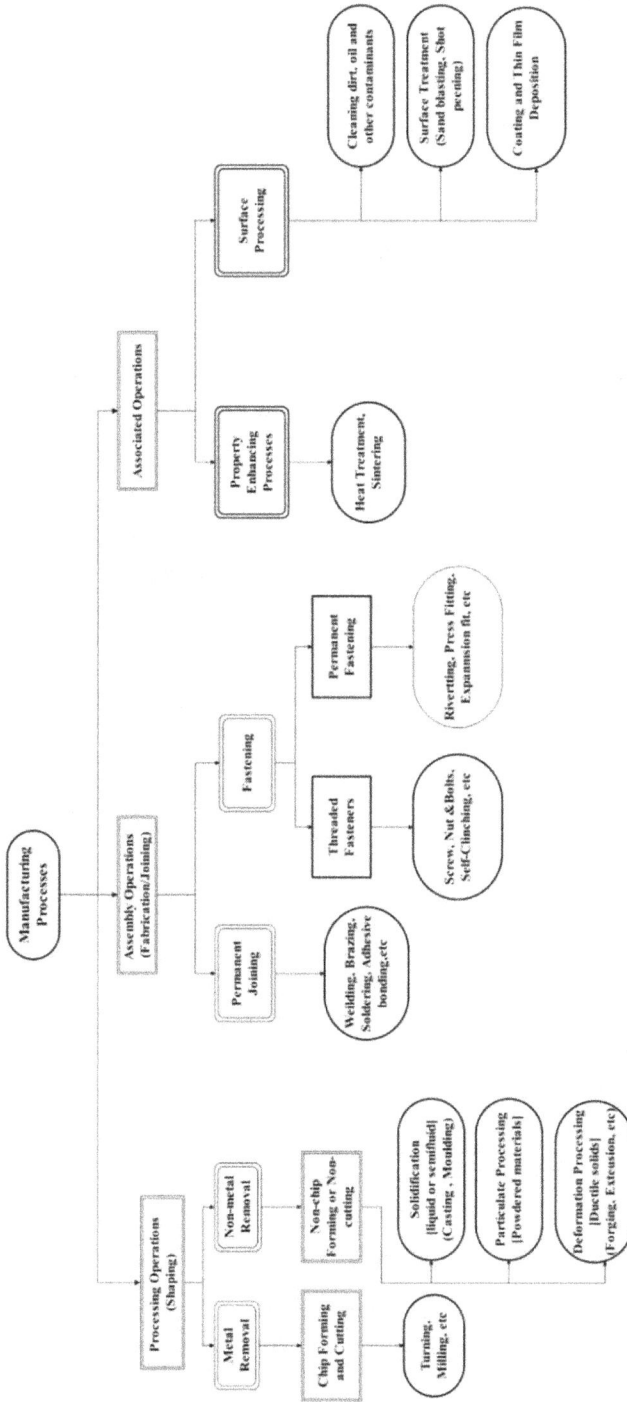

FIGURE 9.2 Classification of manufacturing process.

operation is associated with the shaping of the raw material using various methods. The shaping of the raw material can be achieved through two methods:

a. **Material Removal (Machining Process)**: This process involves the removal of excess materials from the workpiece. An external force is applied to give the desired shape and size to the product. The removal of excess material is aided using different sharp tools harder than the workpiece. The resultant material is removed as chips. In addition, there are two sorts of material removal processes i.e., **conventional machining** (turning, drilling, milling, etc.) and **unconventional machining** (AJM, USM, CHM, etc.), which will be discussed in section 1.4 and 1.5, respectively.

b. **Nonmaterial Removal Process**: This process involves the shaping of materials by heating the materials into liquid state and then putting those liquefied materials into a cavity of desired shape and size. Later, after cooling, the product of the desired shape and size is produced. The most common methods used are *casting* and *molding.* Other methods are *particulate processing*, which involves pressing and sintering of metal powders or ceramics. In *deformation process* force is directly applied to the heated materials to change their shape and size.

2) **Assembly Process (fabrication/joining)**: As during the processing operations (shaping) individual parts are produced. Hence, assembly operations are used to join the two parts into a new entity either by *permanent joining* or *fastening.* Permanent assembly operations can be achieved through welding, brazing, soldering, and adhesive bonding. However, *mechanical assembly* such as reverting, press fitting, nut and bolts are used where the joints are semi-permanent.

3) **Associated Operations**: After the shaping and joining process, there are several associated operations. These operations are used to improve the physical properties of the product. The *property enhancing* process involves heat treatment, which exceeds the physical properties of the manufactured product. In addition, the *surface enhancing* process removes the contaminants from the product's surface via surface treatment, coating, cleaning, and thin film deposition.

Thus, the workpiece undergoes shaping (machining), joining, and finishing processes. This thereby increases its utilization as well as economic value.

9.3 MACHINING PROCESS AND ITS TYPES

The machining process is one of the major reasons for the development of the manufacturing industry. The machining or material removal process involves the removal of material from a workpiece to create a desired product. The removal of materials is achieved through sharp tools and cutting the workpiece in the form of chips. The importance of the machining process can be understood through the factors as given below:

- **Various material workpiece**: A wide range of materials from metals to plastics, glass to ceramics can be successfully shaped.
- **Various geometrical features and component shapes**: It enables us to fabricate and manufacture various designs of different shapes and sizes.
- **Precise dimensional parameters**: It provides dimensional accuracy very close to the desired shape and size.
- **Excellent finished surfaces**: A very smooth surface can be achieved through various machining processes.

However, the wastage of materials (in the form of chips) and time-consuming processes are disadvantages of the machining process.
→ There are two different machining techniques:

1. Conventional Machining Process
2. Unconventional Machining Process (UCM)9.1

9.4   CONVENTIONAL MACHINING PROCESS

Since the time when man discovered how to carve wood and chip stones to create tools for farming and hunting, material removal has been employed for shaping materials. The term "conventional machining process" refers to machining processes that are carried out utilizing only traditional methods, without any cutting-edge techniques. As a result, the term "traditional machining process" is also used to describe this procedure. Sharp point cutting tools, such as the tapping tool in the lathe that is used for tapering, are employed in this operation to get the desired machining results. More damage to the tool happens because the cutting tool is in direct contact with the object being cut and is made of a material tougher than the workpiece. To remove material, a knife is used against a rotating or still workpiece.

Direct interaction between the cutting tool and the object being machined is necessary for the traditional machining process to occur. Metal may be chipped by enabling relative motion between its surface and the cutting device's hard edge. This relative motion is created by the workpiece, the cutting device, or both moving in a mix of rotating and translating motions. The process of cutting metal is referred to as either turning, planning, boring, etc., depending on the characteristics of the relative motion. Multiple kinds of machine tools are required for various processes. A lathe is used for turning, a planer is used for planning, a grinder is used for grinding, etc. Lathes, boring machines, and drills, for example, produce surfaces of rotation, whereas shapers, milling machines, and planers create prismatic (or flat surfaces) components. There are several types of conventional machining processes as shown in Figure 9.3.

TABLE 9.1

Difference between Conventional and UCM

Conventional Machining Process	Unconventional Machining Process
Formation of macroscopic chips due to shear deformation.	Either microscopic or no chip formation takes place.
The real/physical tools are possibly involved during machining process.	Real/physical tools may not be involved.
The cutting tool and workpiece are in direct contact during machining operation.	Neither the cutting tool nor the workpiece is in contact with each other when machining.
The workpiece must not be harder than the cutting tools involved.	The workpiece can be harder than cutting tool.
Cutting tool lifespan is short.	The life of cutting tool is high.
The energy domain can be classified as mechanical.	The energy domain can vary with varying processes such as mechanical, thermal, electrochemical, etc.
Due to application of cutting forces the removal of material takes place.	Need not apply cutting force for material removal.
Provides low accuracy and surface finishing.	Provides high accuracy and enhanced surface finishing.
Every type of materials can be machined economically.	Only certain types of materials can be economically machined.
Wastage of materials is high.	Low wastage of materials.
Causes noise pollution.	Quite process, causes almost no noise pollution.
The equipments are not expensive and easy to setup.	Equipments are expensive and complex to set up.
Both skilled and unskilled persons can operate the cutting tools/machines.	Only a skilled person can operate the machining process.
Generally, manually operated machines are used.	Generally, fully automated machines are used.
Very efficient prototype parts cannot be manufactured.	Very efficient prototype parts can be manufactured.
Hard materials such as ceramic titanium cannot be machined.	Hard materials including ceramic–titanium can be machined.
Mass production can be possible at low cost.	Mass production cannot be possible at low cost.
Material removal rate (MRR) is high.	Material removal rate (MRR) is low.
Lathe machining, drilling, milling, etc. are few conventional machining processes.	Abrasive Jet Machining, Ultrasonic Machining, Electric Discharge Machining, etc. are few UCM.

9.4.1 CHIP FORMATION

The conventional machining process takes place when the tool and workpiece are in physical contact with one another. Chip creation occurs when the material is removed mechanically using instruments like milling cutters, lathes, and saws; particularly when cutting metal with modern, powerful tools made of high-speed steel. The material front of the tool face (also known as the rake face) is squeezed (first elastically, then plastically) as a result of the relative movement that occurs between

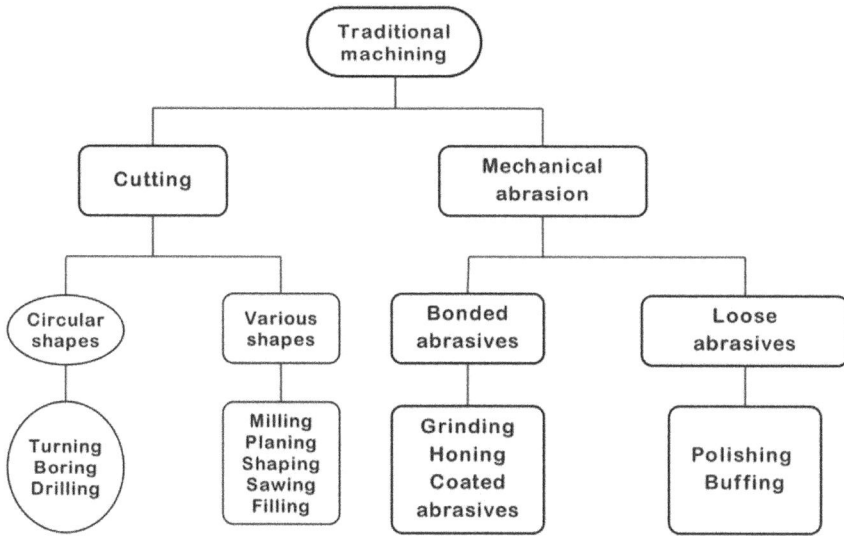

FIGURE 9.3 Classification of conventional machining process.

the tool and the workpiece, as depicted in Figure 9.4. The work material is further distorted plastically as the tool moves into the workpiece, and it is ultimately separated from the workpiece by the deformed material. This separated material is called "chip." Chip curl refers to the curve of the chip, which results when the chip toward the end of the rake face is pulled away from the tool.

9.4.1.1 Types of Chips

Various types of chips are formed while machining, which is dependent on the materials as well as on the angle between the tool and workpiece Figures 9.5 (a) and (b)).

1) **Type I Chip**: Type I chips are created when a tool's upward wedging movement exceeds the material's tensile strength, causing the material to split before the cutting edge and perpendicular to the surface. They are crucial in fibrous materials in particular because of this. Tools with shallow cutting angles produce Type I chips when they are sliced. The size of Type I chips is simply constrained by the length of the cut, which might result in lengthy, continuous chips.

2) **Type II Chip**: Type II chips are produced when a shear force is produced by the wedge-shaped cutter corners. The material warps, creating chips that spiral upward.

3) **Type III Chip**: Type III chips are a better version of Type II chips.

4) **Continuous Chips**: Continuous chips are created in the machining process when ductile material is machined at high speeds with little tool-to-material friction. A ribbon-like continuous chip is produced when ductile materials are cut quickly with a single-point cutting instrument. A continuous chip

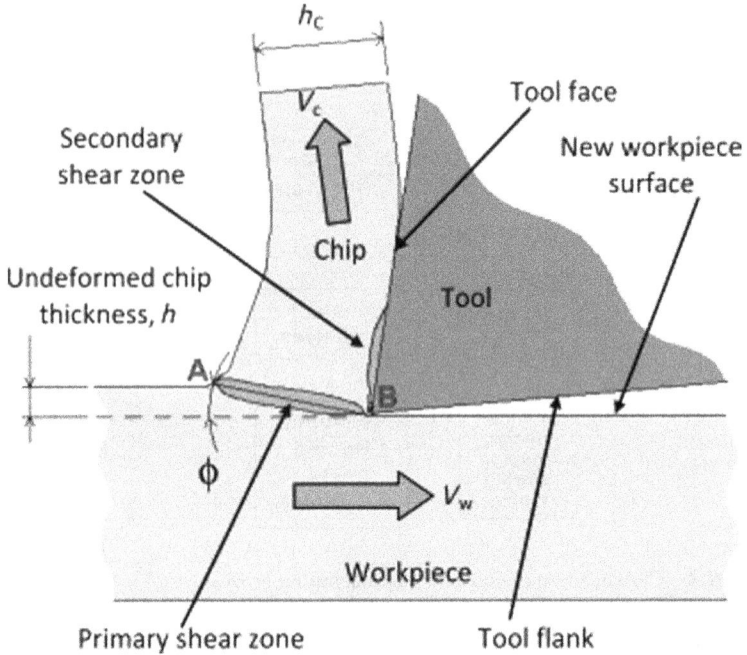

FIGURE 9.4   Chips formation (Haider & Hashmi, 2014).

produces a superior surface polish, a long tool life, and less power usage. However, disposing of big coiled chips, which are generated in tons per week in many businesses, is a significant issue. Various sorts of chip break-ers are employed to solve this issue.

5) **Discontinuous Chips**: Discontinuous chips refer to chips created during the machining process that are broken during cutting or generated in a lim-ited area. Cutting brittle or ductile materials at low speeds under highly frictional circumstances results in the production of segmented or discon-tinuous chips. The tool surface will not be under pressure from the chip since it is broken as soon as it forms, and the blade may simply flatten the remaining uneven surface. These chips offer the workpiece a superior sur-face quality and are shorter in length, making them easier for disposal.

6) **Continuous Chips with Built-Up Edges**: When the temperature, pressure, and friction of the chip against the tool face are high during the machining of ductile material, continuous chips with built-up edges are produced. It is almost identical to the continuous chips, because of the built-up edge and it is blunt.

9.4.2   CLASSIFICATION OF CONVENTIONAL MACHINING

Depending on the process of removal of materials (cutting or abrasion), conventional machining processes are classified into various types, which are discussed below.

(a) Type I: θ = 0° (180°)

(b) Type II: θ = 0c, negative rake angle

(c) Type III: θ = 45°

(d) Type III: θ = 45°, negative rake angle

(e) Type IV: θ = 90°

(f) Type V: θ = 135°

(a)

continuous chip formation

lamellar chip formation

segmented chip formation

discontinuous chip formation

(b)

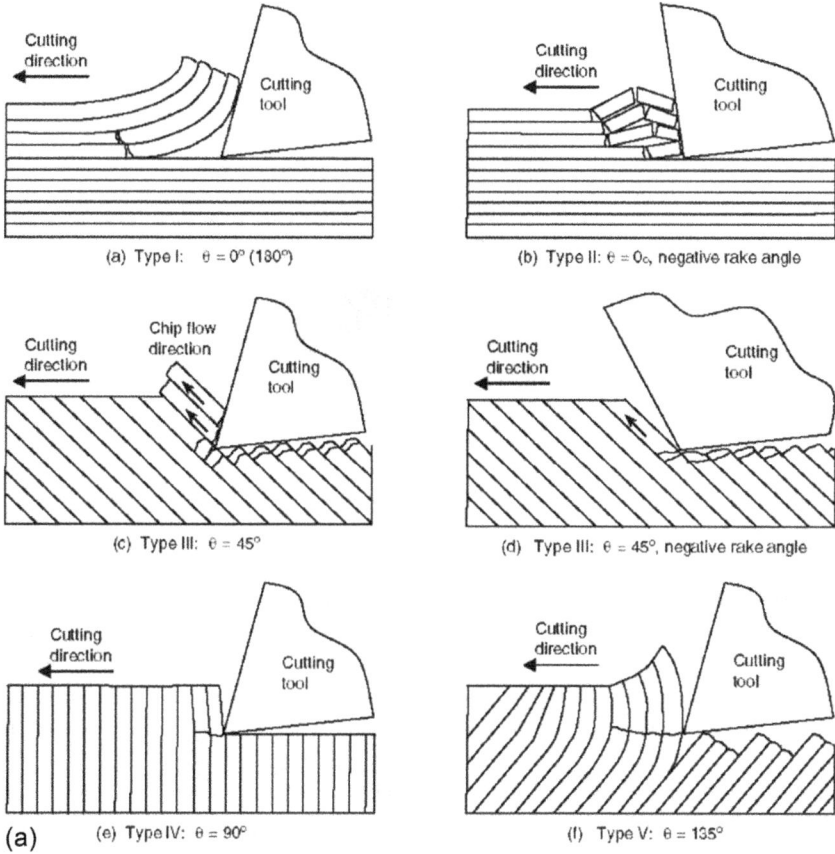

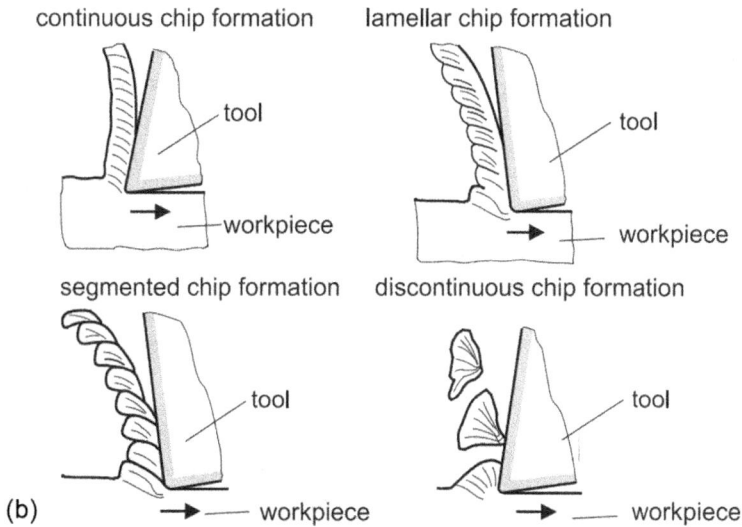

FIGURE 9.5 (a) (Nguyen-Dinh et al., 2018) and (b) (Toenshoff & Denkena, 2013): Types of chips formed during machining.

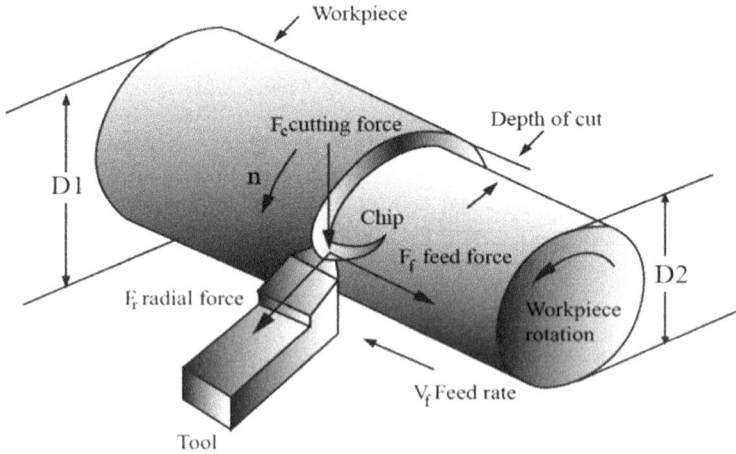

FIGURE 9.6   Schematic diagram of turning process (Alajmi & Almeshal, 2021).

9.4.2.1   Turning

Turning is a type of machining where the material is removed from the surface of a spinning workpiece using a single-point tool. As shown in Figure 9.6, a cylindrical shape is produced by feeding the tool linearly in a direction parallel to the axis of rotation. Utilizing a single-point cutting tool, turning operations are a type of machining that is used to create round objects. To create a complicated rotating shape, materials are removed either by traversing along a predetermined path or in a direction parallel to the axis of rotation. The tool is either connected linearly perpendicularly or parallel to the axis of rotation. In order to produce varied shapes, the cutting tool can be tilted at various angles. To produce various shapes, the cutting tool can be oriented in different ways. Either a CNC turning machine or a manual process might be used. CNC machining is often utilized when highly precise part dimensions are required. It is used to create rotatable, axisymmetric elements like holes, grooves, threads, tapers, various diameter steps, and even curved surfaces. In addition, shafts, engine parts, and parts of machines are all turned.

Turning is often done using a machine tool called a lathe. The lathe provides power to turn the material at a specific rotational speed while moving the cutting tool linearly across the spinning workpiece's surface. Material is chipped away from the entire perimeter until the required diameter is reached. The machining of ductile materials results in the production of continuous chips with built-up edges, when the cutting temperature, cutting pressure, and friction between the chip and tool surface are sufficiently high. The built-up edge makes it abrasive, while being nearly identical to continuous chips.

9.4.2.2   Drilling

In practically all manufacturing facilities and tool rooms, drilling machines are the most basic, versatile, and precise machine tools. Since drilling primarily serves to create holes in the workpiece, it might be considered a single-purpose machine tool. In the process of drilling, a drill is used as a cutting instrument to create circular holes of various sizes and depths. A revolving cylindrical tool with two cutting blades on its working end is typically used for drilling. The instrument is known as a drill or drill bit, as shown in Figure 9.7. A hole with a diameter equal to the revolving drill's diameter is created when it feeds into a stationary workpiece. Twist drills are the most popular drilling instrument and come in a range of diameters and lengths. Drilling machines are used to drill holes, tap, counter bore, ream, and perform other boring operations. Almost all industries employ drilling equipment to create the necessary holes in the workpiece. Additionally, this is employed in carpentry tasks to drill holes in wood and fix wooden constructions.

9.4.2.3 Milling

The most popular kind of machining is milling, which involves removing undesirable material from a workpiece in order to produce a range of characteristics. During milling, an object must be fed against a spinning cutter with several pairs of cutting blades in order to remove metal. The workpiece, which is made of pre-shaped material, is fastened to a fixture that is fastened to a platform within the milling machine. The milling machine also has a cutting tool that revolves at a fast speed and has

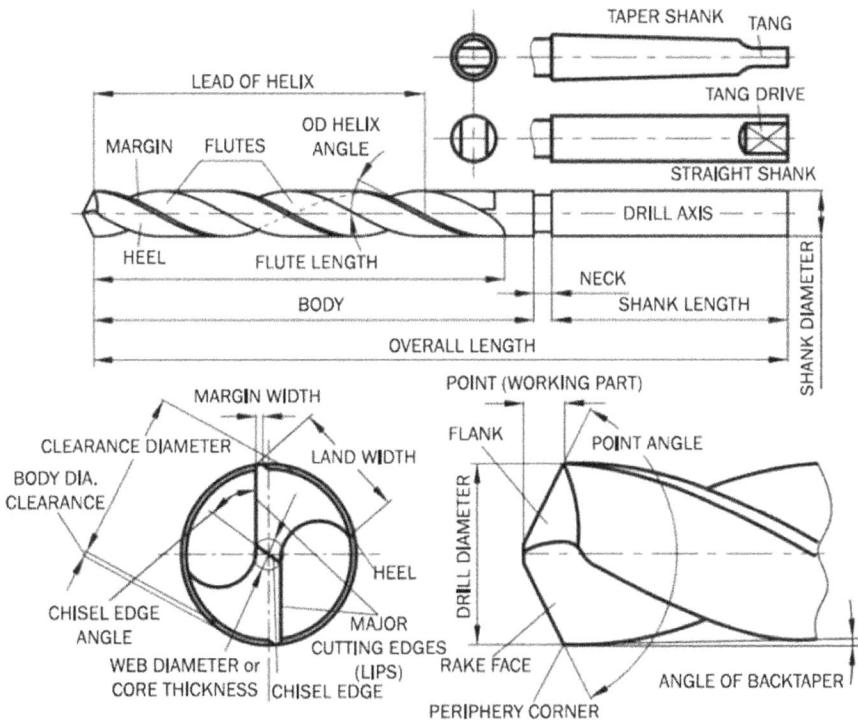

FIGURE 9.7 Illustrations of drill bit (Astakhov, 2011).

sharp teeth. When a workpiece is fed into a spinning cutter, the material gets carried away as little chips to give the workpiece the ideal shape. The flat or curved surfaces of a variety of workpieces may be accurately and nicely finished using milling machines. Using the appropriate attachments, a milling machine may also be used for drilling, slotting, forming a circular profile, and gear cutting.

In Figure 9.8, two fundamental categories of milling processes are depicted.

1) **Peripheral Milling**: In plain milling, commonly referred to as peripheral milling, the cutting edges on the cutter's outside perimeter execute the action while the tool's axis is parallel to the surface being machined. Two types of milling, up milling and down milling, are distinguished in peripheral milling by the direction of cutter rotation. Up milling, also known as

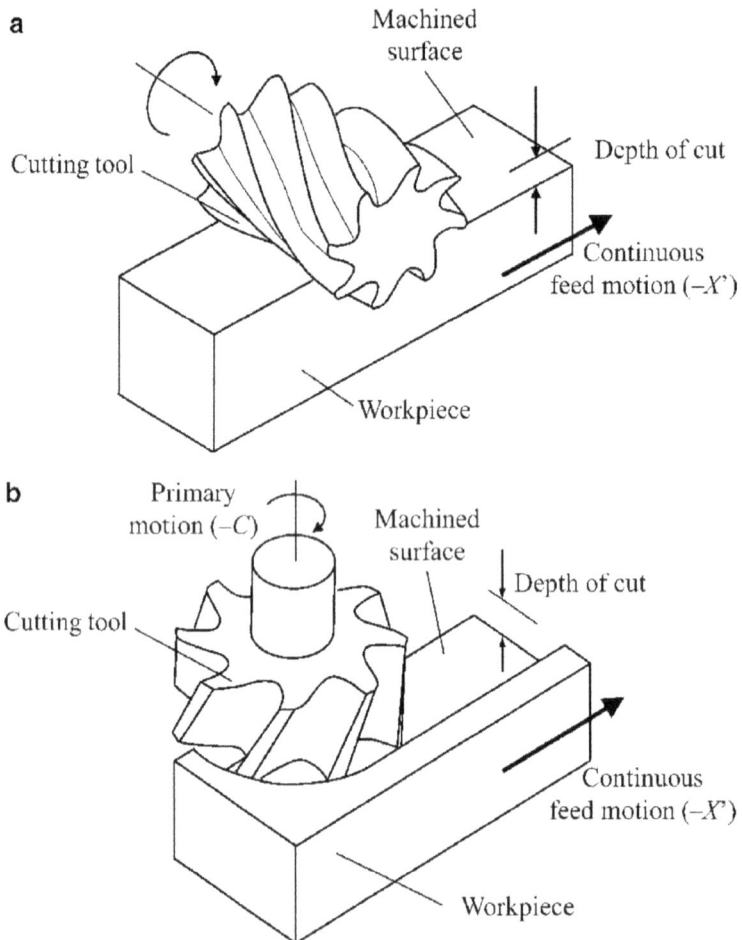

FIGURE 9.8 Different types of milling process (Girsang & Dhupia, 2015).

conventional milling, is a cutting process where the cutter teeth move in the opposite direction from the feed direction. The milling is done "against the feed." The cutter moves in the same direction as the feed when cutting the work in down milling, also known as climb milling.

2) **Face Milling**: In face milling, the axis of cutter is parallel to the surface that will be cut. Face milling creates smooth surfaces and flat-bottomed vacancies in the workpiece. For example, ball-nose cutters can be used to mill pockets with curved walls. Special cutters are available for a range of jobs.

9.4.2.4   Grinding

The most popular kind of abrasive machining is grinding. Small quantities of material are removed by grinding from both cylindrical and flat surfaces. Cutting is an aspect of grinding since grinding cuts metal. Surface grinders feed the work through the grinding wheel in a circular motion. The wheel typically cuts between 0.00025 and 0.001 inches deep. A grinding wheel or grinder is the cutting tool used in the abrasive machining process of grinding. It is a method of cutting materials that makes use of an abrasive instrument with grains that are similar to grit, as shown in Figure 9.9. Sharp cutting tips, high hot hardness, chemically resistant, and wear resistance are the characteristics of these grits. An appropriate bonding substance holds the grits together to form the shape of an abrasive tool. High precision and excellent surface quality are obtained on the workpiece by using a grinding machine. The cement industry and mineral processing facilities frequently use grinding.

9.4.3   ADVANTAGES OF CONVENTIONAL MACHINING PROCESS

1) **Cost-effectiveness**: Conventional manufacturing methods are frequently less expensive than their technologically advanced equivalents because they use less high-tech devices and automated systems.

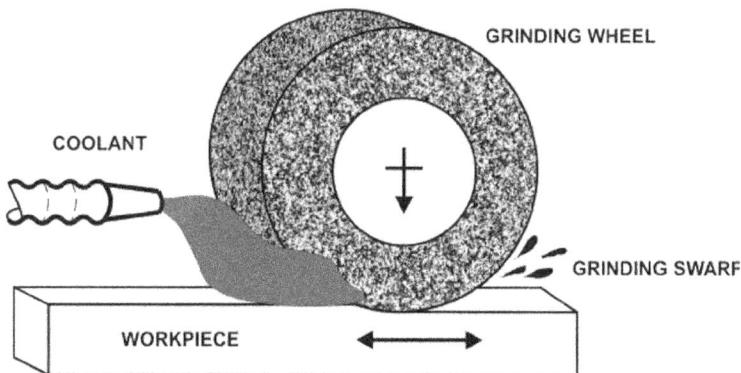

FIGURE 9.9   Schematic diagram of a grinding machine (Rubio & Jáuregui-Correa, 2012).

2) **High Modification**: Traditional manufacturing techniques allow for high levels of customization, which makes it simpler to develop goods that are specifically suited to consumer needs.

3) **Specialized Labor**: Skilled personnel are frequently required in traditional manufacturing processes to operate machines, do assembly and inspection work, among other activities.

4) **Reliability**: Because conventional manufacturing techniques have been in use for so long, it is clear that they are reliable and stable.

9.4.4 LIMITATIONS OF CONVENTIONAL MACHINING PROCESS

1) **Low Efficiency**: Traditional manufacturing methods are frequently less effective than modern ones.

2) **Limited Output Capabilities**: Traditional manufacturing techniques sometimes have low output levels and can fail to create sophisticated or complicated goods.

3) **High-skilled Labor**: Skilled labor is frequently needed, which can be physically taxing and time-consuming. Increased manufacturing costs and worker health and safety issues may come from this.

4) **Challenges with Quality Control**: Human mistakes may be more likely to occur through traditional production processes, creating difficulties with quality control.

5) **Technological Obsolescence**: Traditional production methods sometimes rely on outmoded, older technologies.

9.5 UN-CONVENTIONAL MACHINING PROCESS

The conventional machining process became outdated with the rise of modern industries. It cannot be used to machine the newly developed alloys efficiently. Also, the conventional machining process is not sustainably beneficial, economical, and efficient. Consequently, the development and rising demand of newly developed "hard to machine" materials including ceramic, hard steel alloys, tungsten, stainless steel etc. Due to this, the scientists and engineers were compelled to develop a novel UCM process.

The UCM is not dependent on sharp tools for the removal of excess materials. Rather, it utilizes alternative tools that use various energy sources for machining actions. The source of energy involved can be mechanical, thermal, chemical, and electrochemical in nature, as shown in Figure 9.10. Because of the involvement of such a vast energy source the UCM enables the manufacturers to work on any type of workpiece, whether it is hard, brittle, soft, or flexible. It also provides an ease of fabricating complex shapes and structures, which cannot be obtained through conventional machining.

9.5.1 CLASSIFICATION OF UCM PROCESSES

As discussed above, the UCM process utilizes various energy sources for shaping of material. Thus, on the basis of energy source involved the UCM can be divided into four categories, as shown in Figure 9.10.

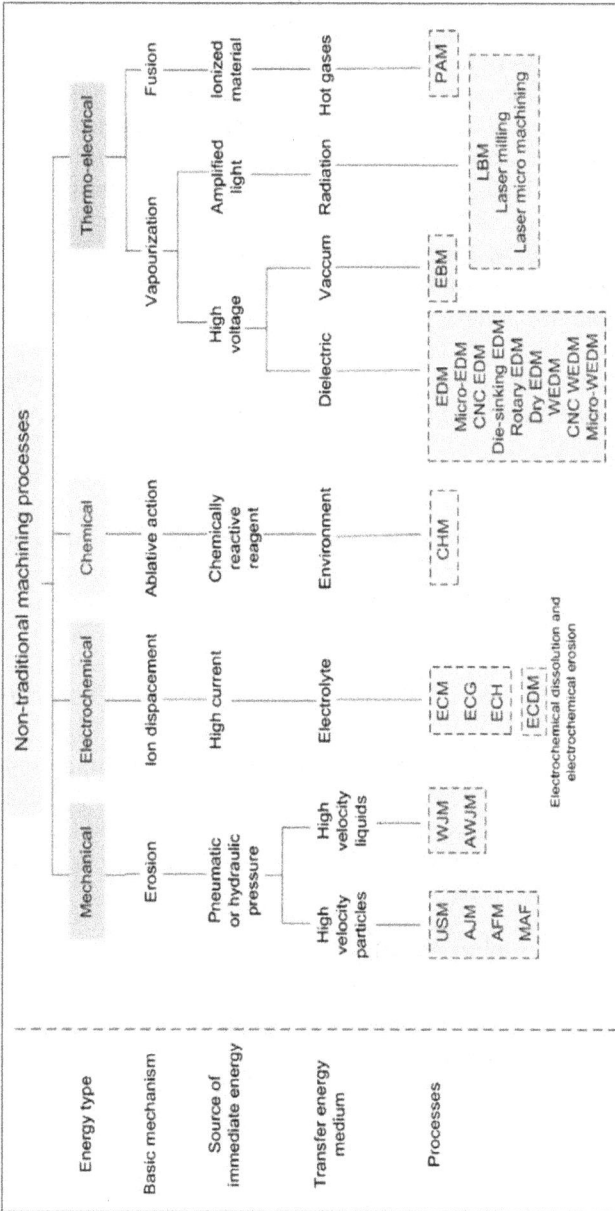

FIGURE 9.10 Classification of UCM (Prasad & Chakraborty, 2018).

1) **Mechanical Energy Based**: This process utilizes mechanical force for the removal of workpiece material. Mechanical force indicates that tools or methods used in this process use kinetic energy or potential energy. In most cases kinetic energy is used to remove the materials. Water Jet Machining (WJM), Ultrasonic Machining (USM), Abrasive Jet Machining (AJM) etc. uses mechanical force to remove excess materials.

2) **Chemical Process**: In the chemical process the material (mainly metals) is eliminated from the piece of metal by carefully regulated etching or chemical-based attack. Certain types of acids or other etchants are used to remove unwanted material. The selective removal of material is achieved by applying a mask on the surface of the material that is not to be etched. The processes involved are chemical and photo-chemical milling.

3) **Electrochemical Energy Based**: This process is a reverse process of electroplating but with some modification. In this process, the removal of material is achieved through displacement of ions. Through the space between the tool and the workpiece, an electrolyte circulates. The processes involved are Electrochemical Machining (ECM), Electrochemical Grinding (ECG) etc.

4) **Thermal Energy Based**: In this process heat is applied for the removal of layers of work surface. The erosion of materials takes place by fusion or vaporization. Mainly the heat is produced by the conversion of electric energy. Electron Beam Machining (EBM), Laser Beam Machining (LBM), and Plasma Arc Machining (PAM) are a few methods that use thermal energy for erosion.

9.5.2 MECHANICAL ENERGY-BASED UCM

Mechanical energy-based UCM processes work on the process of erosion of workpiece material utilizing pressure. The machines either work on pneumatic (air or gas) or hydraulic (liquid) pressure. The machining medium can be either particles suspended in air and fluid or only fluid. Some of the mechanical energy-based UCM are discussed below.

9.5.2.1 Water Jet Machining (WJM)

Water Jet Machining (WJM) or hydro-dynamic machining uses highly pressurized water as a medium for cutting the workpiece. The highly pressurized water striking the workpiece cut the workpiece into desired shapes, as shown in Figure 9.11. The WJM is mainly used to cut soft materials like wood, paper, leather, textiles, polymers etc. As WJM uses only water as a source for cutting, it does not involve production of gases or dust particles during machining. WJM also does not cause pollution to the environment. One of the main advantages of using WJM is that it does not thermally damage the products, as no heat is involved during cutting. Also, the machining surfaces are clean, polished, and a minimum burr is present. Materials irrespective of their physical and chemical properties can be machined using WJM. As WJM involves no cutting tools the maintenance of tools is easy, although other parameters

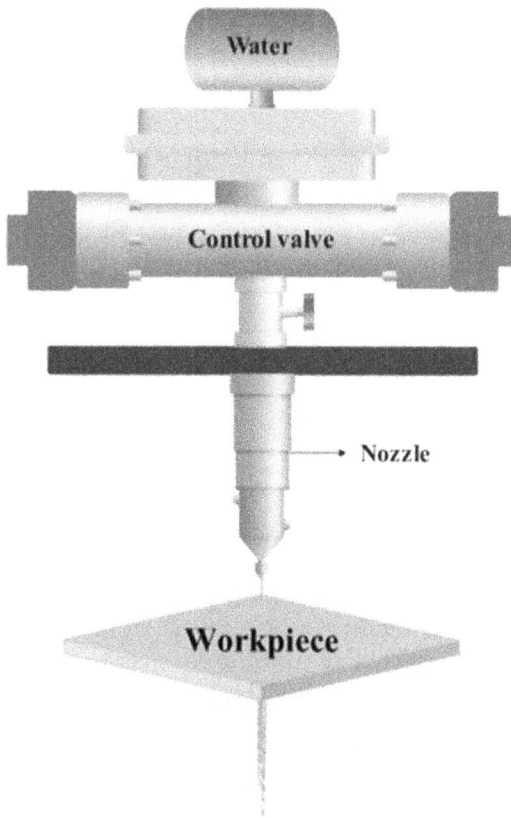

FIGURE 9.11   Schematic diagram of WJM.

should be controlled carefully, this regulates the WJM working process, as shown in Figure 9.12. WJM is widely used in automotive, food, aerospace, defense, medical, and mining industries. In automotive and aerospace industries WJM is used for drilling, grooving, cutting etc.

9.5.2.2   Abrasive Water Jet Machining (AWJM)

Water Jet Machining is limited to cutting soft materials. However, hard and brittle materials such as ceramic, glass, metal composites need to be processed using AWJM. AWJM is the improved version of WJM, first developed in 1974. The addition of abrasive particles such as silicon carbide or aluminum oxide, the WJM process not only enhances the material removal rate (MRR) but also enables cutting through hard materials (Bhowmik et al., 2019). Figure 9.13 depicts the AWJM basic diagram.

AWJM has two components involved in the cutting process i.e., fluid (water) and abrasive particles. Both water and abrasive particles complement each other, where abrasive particles increase the erosion of materials, whereas water acts as a carrier

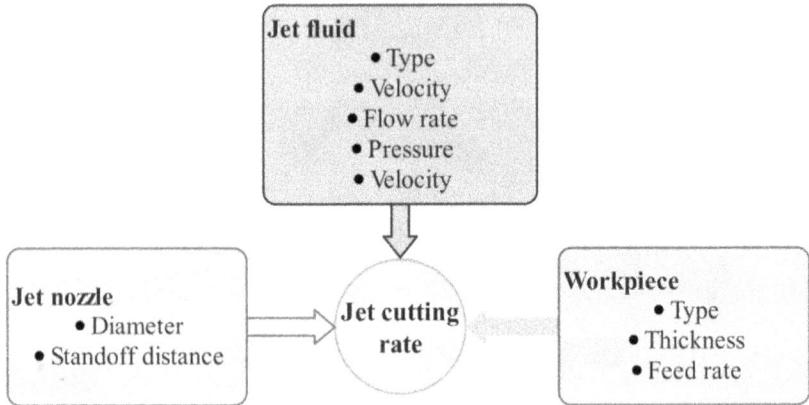

FIGURE 9.12 Factors affecting the WJM jet cutting rate (Authors).

for both abrasive and eroded materials and also acts as a coolant. AWJM enables to cut parts accurately and efficiently. AWJM is ideal for several materials which are sensitive to heat and which cannot be machined using laser or thermal cut. Thus, AWJM is used in the manufacturing industry for cutting parts for automobiles, ships, and aircraft. Further, it is used in the construction industry to cut ceramic tiles and marbles. In addition, the electronic circuit boards are also machined using AWJM.

9.5.2.3 Abrasive Jet Machining (AJM)

In AWJM, the abrasive particles are assisted by high-pressure liquid or water but in AJM the abrasive particles are assisted by fast-flowing gases at several atmospheric pressures. This technique of impact erosion removes materials off the surface by employing a focused fast-moving stream of granular abrasives enclosed in a fast-moving gas jet. The finer particle size and versatile machining settings set it apart from traditional sandblasting. In AJM the impingement of abrasive particles leads to the removal of materials. High-velocity abrasive particles impinge the workpiece with great force, which causes brittle fracture in a workpiece. The repetitive striking of abrasive materials leads to the removal of fractured materials with fast-blowing air/gas leading to cavity in the workpiece. The Mechanical Abrasion (MA) operation produced by fast-moving abrasive particles removes the workpiece material, as shown in Figure 9.14. The best applications for AJM machining include making holes in extremely hard materials. Typically, it is employed in cutting, cleaning, penning, and etching hard materials like ceramics, metals, and glass.

The abrasives used are aluminum oxide, silicon carbide, and glass beads (grains with size of 10–50 μm). The shape of abrasive particles may be irregular or spherical (sharp-shaped abrasive particles are taken because of good wear action).

The variables that affect the cutting phenomena are:

1) *Abrasive*: composition, strength, shape, size, and mass flow rate.
2) *Carrier gas*: composition, pressure, and velocity.

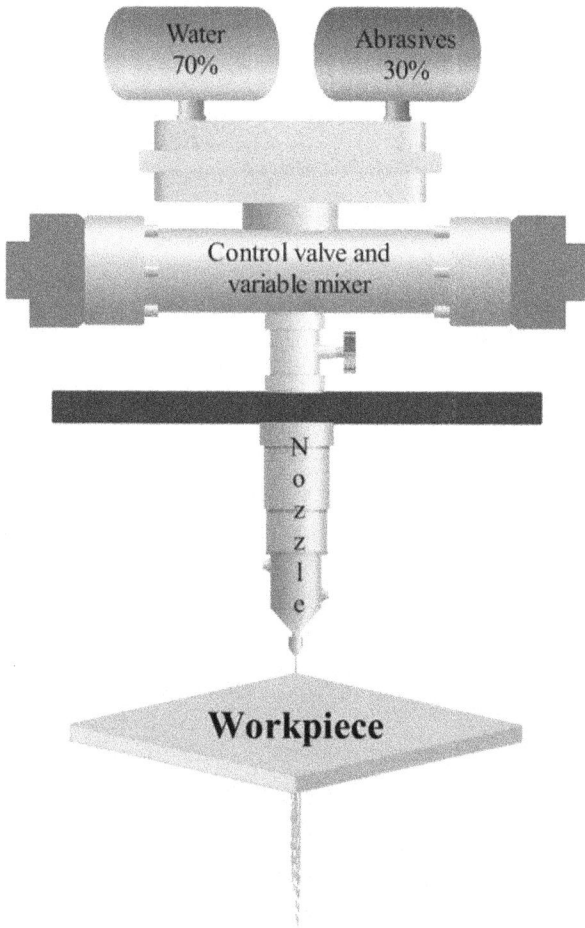

FIGURE 9.13 Components of AWJM process.

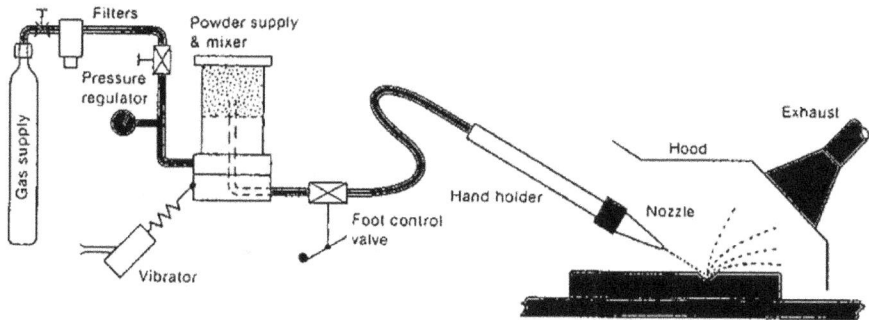

FIGURE 9.14 Setup of AJM (Srikanth & Rao, 2014).

3) *Nozzle*: geometry, composition, nozzle-tip distance (stand-off distance and its orientation).

9.5.2.4   Ultrasonic Machining (USM)

Similar to AWJM, USM also uses abrasive particles suspended in fluid for erosion of materials. However, in USM abrasive slurry and vibrating tools in different shapes are used for material removal process. The first USM technology was developed in 1953 and 1954 and mounted on milling and drilling machine components. Subsequently, in 1960, standalone USM machines were employed for routine manufacturing in several kinds of implementations. USM is a kind of nontraditional machining technique which utilizes the vibrating tool to induce removal through mechanical processes, aided with abrasive substances, such as slurry and liquid, in between both the workpiece and the tool. The mixture of fine abrasive particles, including boron carbide, aluminum oxide, and silicon carbide, that make up the abrasive slurry comports with a fluid medium (Daud et al., 2022). Because USM operations are independent of the material's electrical or chemical properties, they can be used to machine a wide range of materials. Thus, this method is frequently employed in the machining of brittle and hard materials including Si, quartz, borosilicate glass, titanium alloys, and ceramics that are challenging to work with using conventional methods.

Figure 9.15 shows the basic configuration for the USM process. The main components of the setup are a tool system, which employs a transducer that converts electrical energy into mechanical energy, an ultrasound frequency vibration tool with an operating frequency range of 20–40 kHz, and a slurry feed unit (Daud et al., 2022). The cutting zone between the vibrating tool and the workpiece is continually fed with the abrasive slurry of B_4C or SiC during the oscillation. The vibration causes the slurry's abrasive particles to smash against the fixed workpiece, creating microscopic indentations that start to fracture the material and chip off tiny pieces of it. The oscillating tool applies a static pressure to the abrasive grains with amplitudes

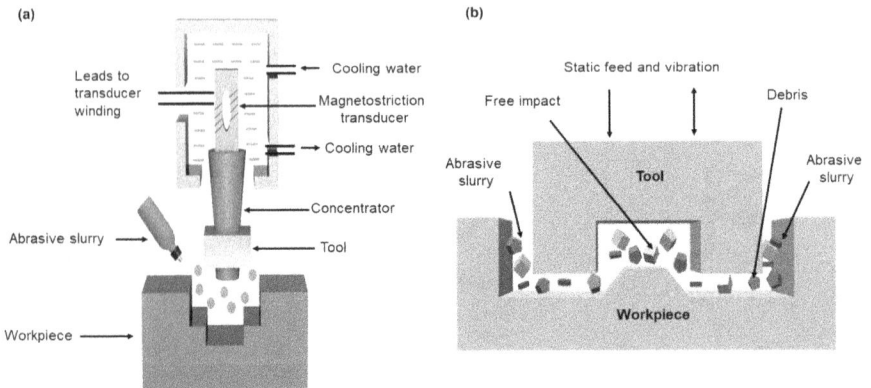

FIGURE 9.15   Setup for USM (Daud et al., 2022).

between 10 μm and 40 μm and goes down while the material breaks down to generate the necessary tool shape.

In the USM operation, there are typically four different kinds of material removal mechanisms: mechanical abrasion, micro-chipping, cavitation effects, and chemical reactions. When using a mechanical abrasion mechanism, the abrasive particles are directly hammered on the surface of the workpiece to remove minute amounts of material, as shown in Figure 9.16 (a). In contrast, micro-chipping can be done by oscillating the machine tool, which allows the abrasive slurry to circulate freely and has an impact on the workpiece's surface. The collapse of gas bubbles in liquids that have been triggered by ultrasonic vibration, known as cavitation effects, can also result in the material removal process. The presence of chemical pollutants in the

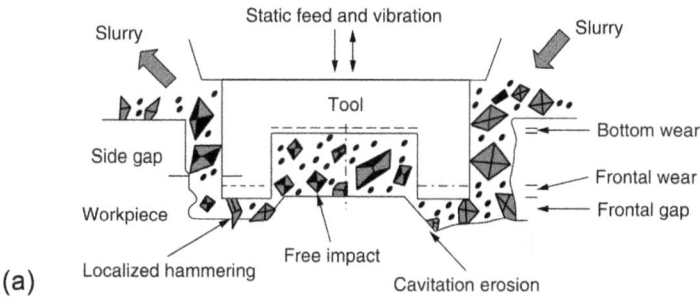

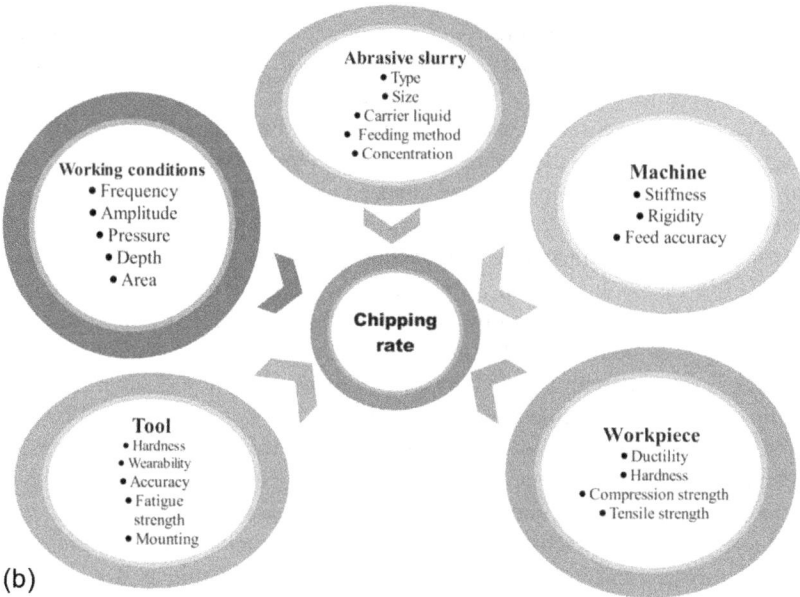

FIGURE 9.16   (a) Working mechanism of USM (Singh et al., 2018). Factors affecting USM Chipping rate (Authors).

slurry, the chemical action degrades the workpiece material, resulting in the loss of material. There are several factors that affect the USM working, as depicted in Figure 9.16 (b).

USM enables exceptionally precise cutting of complex and irregular shapes. Therefore, it is utilized in optical and electrical applications that need more precise and high-quality cutting. In contrast to other metal removal techniques like chemical and thermal ones, USM preserves the crystallographic qualities of the workpiece without affecting its physical characteristics. They are therefore employed in the machining of more fragile and delicate materials. The materials that are most frequently machined using USM include hardened steels, zirconia, glass, ceramics, and precious stones.

9.5.3 CHEMICAL ENERGY-BASED UCM

9.5.3.1 Chemical Machining (CHM)

Chemical Machining (CHM) is considered to be one of the earliest and most effective machining techniques. It involves utilizing a reactive chemical solution, such as a powerful alkaline or acidic reagent, to regulate a chemical reaction that dissolves and removes the material from the workpiece (Bhattacharyya & Doloi, 2020; Çakir et al., 2007). In ancient Egypt, about 2300 BC, citric acid was used as the first chemical machining tool to etch copper jewelry. This procedure was typically employed for decorative etching up until the 19th century (Harris, 1976).

Chemical energy-based machining processes work on the principle of ablation (etching) of a workpiece by means of a chemically reactive agent (etchant). The workpiece is covered with maskants and immersed in etchants such as acids or bases. Maskants are chemically resistant materials (acid or base resistant), which are used to cover the parts of workpiece that are not to be etched (Çakir et al., 2007). Using this method, materials from portions with a high strength-to-weight ratio are removed while also creating gaps and shapes (El-Hofy, n.d.). Additionally, a lot of micro-components are made using the machining technique for usage in a variety of industrial applications, including semiconductor and micro-electromechanical systems (MEMS) technology (Dini, 1984; El-Hofy, n.d.). The removal of materials in chemical machining is obtained by immersing the workpiece in chemical etchants (Bhattacharyya & Doloi, 2020; Raj et al., 2019). The solid material is removed due to the etchant's reaction with the workpiece, as shown in Figure 9.17. There are several steps involved during chemical machining (Bhattacharyya & Doloi, 2020; Çakir et al., 2007; El-Hofy, n.d.; Raj et al., 2019), as shown in Figure 9.18. CM provides a wide range of applications that may be used with tiny to extremely big workpieces. Different types of nonmetal components including glass, plastic, ceramics, etc. are etched using chemical machining, as well as different metals like titanium, copper, nickel, aluminum, and its alloys. With the help of this technique, it is possible to create shallow cavities and holes with a wide surface area, as well as curves, pockets, depressions, and engraving on metallic components. Microelectronic devices are also made using this method. The electronics sector makes extensive use of CM to produce metallic components such as enclosure screens, recording heads, instrument

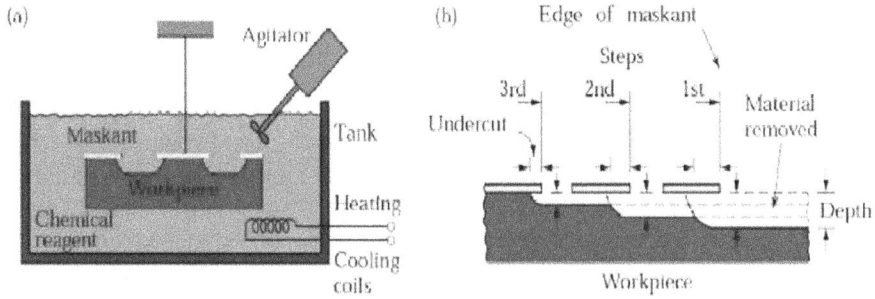

FIGURE 9.17 (a) Process flow diagram for chemical machining, (b) steps involved in creating a profiled cavity by chemical machining (Raj et al., 2019).

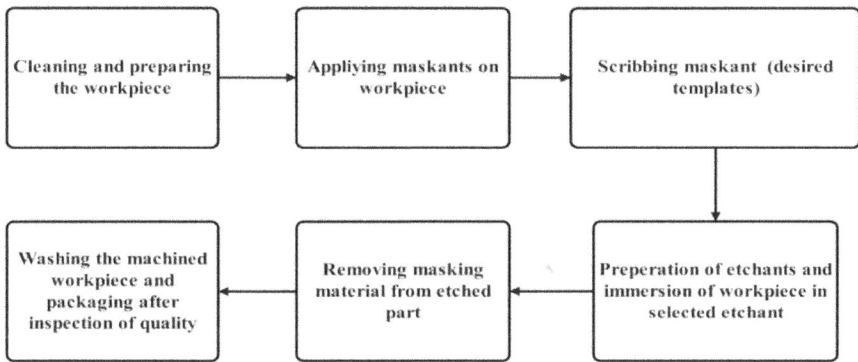

FIGURE 9.18 Process flow of chemical machining process (Authors).

panels, semiconductor devices, and Printed Circuit Boards (PCB), Integrated Circuits (IC), and other circuit devices (Al-Ethari et al., 2013; Allen & Talib, 1983; Black & Kohser, 2020; Drozda & Wick, 1989; Tehrani & Imanian, 2004).

9.5.4 ELECTROCHEMICAL ENERGY BASED UCM

Electrochemical energy-based machining processes work on the principle of reverse electroplating or Electrochemical Dissolution (ECD). In this process, the chemical reaction is initiated in the presence of electric current. Through the anodic dissolving process, Electrochemical Machining (ECM) is used to manufacture objects made of metal and metal alloys regardless of their hardness, strength, or thermal qualities (Pandilov, 2018).

9.5.4.1 Electrochemical Machining (ECM)

The umbrella name for a number of electrochemical methods is Electrochemical Machining (ECM). Regardless of their strength, hardness, or thermal qualities, ECM is used to manufacture components made of metal and metal alloys.

Electrochemical Machining (ECM) is a noncontact, reverse electroplating procedure that removes material rather than depositing it. In ECM, a strong electric current is sent through a conductive fluid from tool to workpiece. The workpiece's metal atoms are ionized and removed using a high current and conductive fluid, producing a surface devoid of burrs (Bhattacharyya & Doloi, 2020). Michael Faraday (1791–1867) developed the electrochemical anodic dissolution theory long ago, in the 19th century. The two fundamental rules of electrolysis that Faraday developed in 1833 serve as the fundamental basis of the electro-deposition and electro-dissolution processes. The first method of machining metal anodically by an electrolytic process was created and patented in 1929 by the Russian researcher W. Gusseff (Bhattacharyya & Doloi, 2020; El-Hofy, n.d.). Given that the method is based on electrolysis (i.e., chemical change, particularly degradation, caused in an electrolyte by an electric current), it is incredibly precise and capable of milling any electrically conducting workpiece. The aerospace, automotive, building, medical, micro-systems, and power supply sectors all employ ECM (Pandilov, 2018). However, the remarkable characteristic of electrochemical machining is that it is suited for cutting any conductive material as well as intricate cavities in high-strength materials since it is independent of the strength, hardness, and toughness of the workpiece material. In particular, high-alloyed nickel- or titanium-based metals and hardened materials may be electrochemically machined, however nearly any type of metal can be done so (Deconinck et al., 2012). Automotive, offshore petroleum, medical engineering, and aerospace companies all used ECM in a variety of machining applications, while aerospace companies continue to be its main user today (McGeough, 1974). ECM has recently drawn a lot of interest in the manufacturing of small components for a number of micro-engineering projects (Taniguchi, 1983).

Regardless of the metal's structural integrity, machining may be done on both soft and hard materials, including rare metals. This technique can be used to make curves, ring ducts, grooves, or bell hollows without any touch and with extreme accuracy (Pandilov, 2018). The fundamental process and key components making up a typical ECM system setup are shown in Figures 9.19 and 9.20, respectively.

9.5.4.2 Types of ECM Process

There are different types of machining processes that utilizes electrochemical machining methods such as:

(1) **Electrochemical Honing**: The removal of material by the use of electrical energy, chemical energy, and the honing process is known as electrochemical honing. To polish the inside surface, electrochemical honing simultaneously uses mechanical abrasion and anodic dissolution. Although it resembles traditional honing, anodic dissolution is used to remove the majority of the metal in this process. Any hard electrically conductive substance can have its inner surface of the hole polished by EC honing.

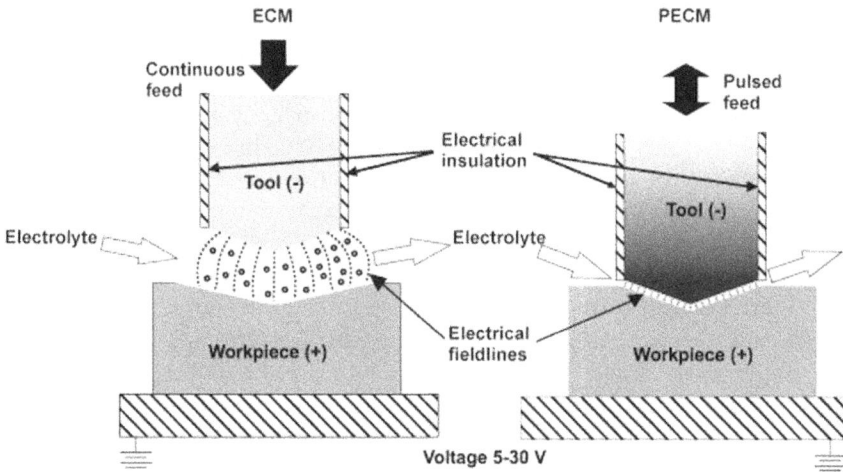

FIGURE 9.19   The fundamentals of precise/pulsed electrochemical machining (PECM) and electrochemical machining (ECM) (Burger et al., 2012).

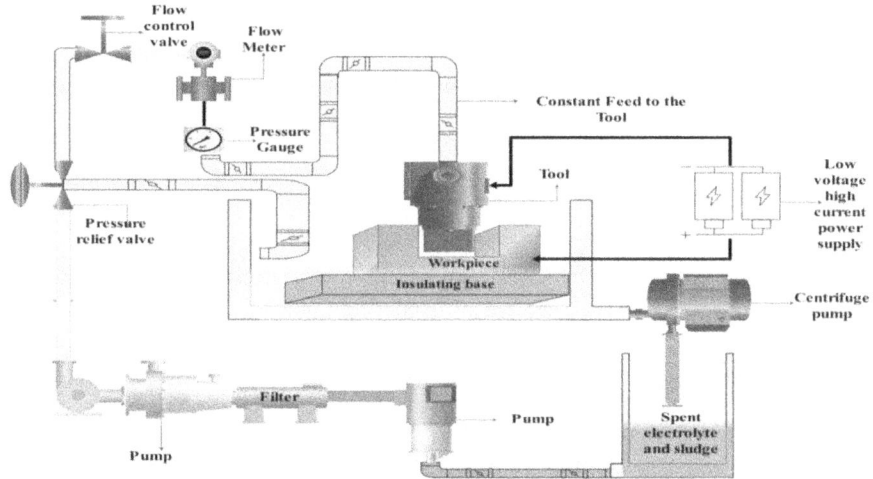

FIGURE 9.20   The various components of apparatus used for electrochemical machining

(2) **Electrochemical Grinding**: The mechanical removal of material employing charged grinding wheels is combined with electrolytic action in electrochemical grinding. This procedure is comparable to mechanical grinding, which produces superior surface finishes. However, a metallic wheel is used in this instance rather than an abrasive attached grinding wheel (Figure 9.20). A little space is kept between the metal wheel and the workpiece, and it is filled with electrolyte solution (Bhattacharyya & Doloi, 2020). Contrary to typical grinding, there is no interaction

between the grinding device and the workpiece; hence there is zero cutting force, heat generation, deformation, or stress development on the object being ground.

(3) **Electrochemical Drilling**: Electrochemical drilling is the name of the process where ECM is simply utilized for drilling holes. When compared to traditional drilling, electrochemical drilling is significantly quicker. A continuous heated electrolyte is fed through the tool and out the tip, where accelerated corrosion from electrochemical machining dissolves the workpiece.

(4) **Electrochemical Deburring**: Deburring is the process of removing burrs from a workpiece in order to create a smooth final surface. Typically, when an object is produced using a traditional machining technique, burrs are left along the two crossing surfaces. Such burrs are unfavorable from the perspectives of a component's performance and an operator's safety. When other deburring techniques are ineffective, electrochemical deburring is typically used for remote and inaccessible locations. In order for the electrochemical reaction to occur, this technique uses a flowing electrolyte as a conductor of electric current.

9.5.5 THERMAL ENERGY-BASED UCM

Thermal energy-based machining processes work on the principle of generating heat through various methods such as electric current, laser, high-energy electrons, and ionized gases. There are various machining processes which use thermal energy for the removal of materials. Some of them are discussed below:

9.5.5.1 Electric Discharge Machining

Priestley made the initial discovery of electrical discharge's corrosive action in 1770. EDM is a significant example of a nontraditional machining technique for forming complicated components made of conductive materials (Lauwers et al., 2004; Petrofes & Gadalla, 1988; Sanchez et al., 2001). EDM may be conceptualized as a thermoelectric process that uses repeated electrical discharge (sparks) to slowly remove electrode and workpiece materials that are electrically conductive (Palanikumar & Davim, 2013). The procedure is a novel machining technique that enables the fabrication of burr-free components with sophisticated shapes irrespective of the hardness of materials (El-Hofy, n.d.; Pedroso et al., 2023). EDM is typically required for manufacturing molds with the proper dimension, depth, and form.

EDM is based on the electro-discharge erosion (EDE) action of electric sparks happening between two electrodes that are kept apart by a dielectric liquid, as shown in Figure 9.21. In EDM, material is removed from the workpiece by passing a pulsating (ON/OFF) high-frequency current via the electrode. By doing this, extremely small fragments of material are gradually removed (eroded) from the workpiece. The high-frequency current forms plasma channels leading to very high temperature and pressure. At such a high pressure and temperature, some metal is melted and eroded (Choudhary & Jadoun, 2014; Palanikumar & Davim, 2013). There is just

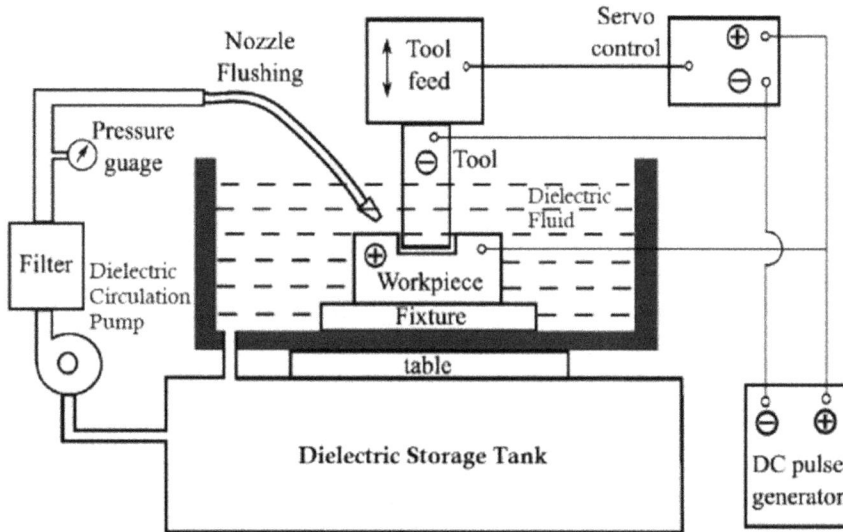

FIGURE 9.21 Principles of EDM (Joshi & Joshi, 2019).

a partial removal of the molten metal. When the potential difference is removed, the plasma channel stops functioning. At such high pressure and temperature, some metal melts and erodes when the plasma channel collapses, and shock waves are created to remove the molten material. Material removal results from intense local temperature rise, as shown in Figure 9.21 and 9.22.

9.5.5.2 Laser Beam Machining (LBM)

The term "laser" stands for "light amplification by stimulated emission of radiation." Laser material processing, often known as LBM, is more widely used to refer to procedures for machining and material processing including thermal treatment, alloying, cladding, metallic sheet bending, etc. This process is conducted using coherent photon or laser energy, which is mainly transformed into thermal energy upon contact with the majority of materials. The exact portion of the metal will be machined depending on the laser beam's course of travel (Basu et al., 2020). Depending on the material, the laser strength can be changed in this situation. The sole attribute necessary for laser beam machining is the metal's laser absorption capacity, or its capacity to spontaneously get heated upon laser beam engagement. In order to avoid oxidation while working with metals, laser beam machining must be done in an inert gas environment. For precision machining complex parts, laser beam machining is primarily used in the automotive, aerospace, shipbuilding, electronics, steel, and medical industries. In heavy manufacturing, it may be used for drilling and cladding, seam, and spot welding, among other things. It is employed in the medical field for cosmetic surgery and hair removal.

The workpiece which needs to be processed is the target of the generated laser beam. The heat energy impinges on the workpiece when the laser hits it. This will

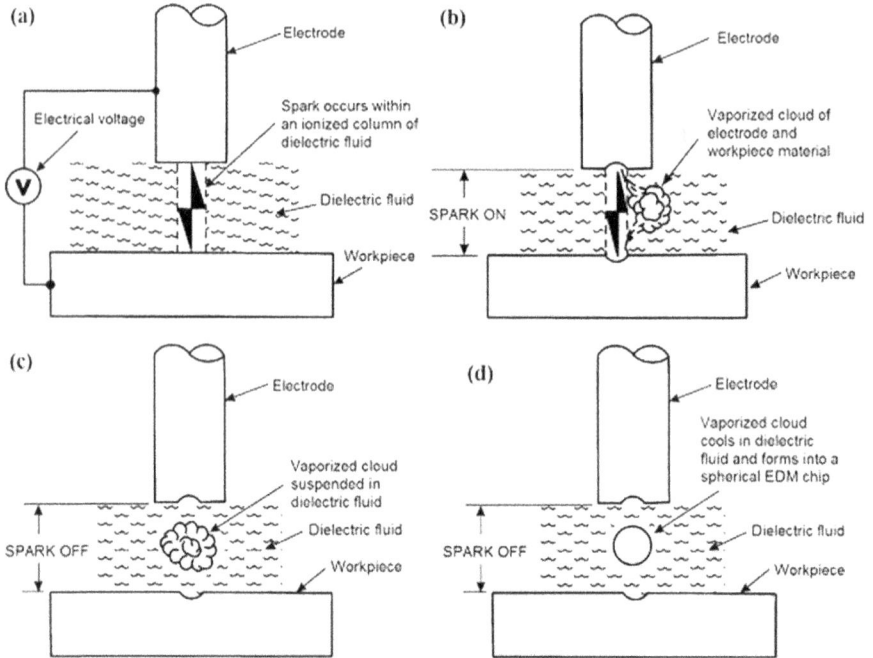

FIGURE 9.22 (a) Spark ON: electrode and workpiece material vaporized. (b) Spark OFF: vaporized cloud suspended in dielectric fluid. (c) Spark-OFF: vaporized cloud solidifies to form EDM chip (debris). (d) Spark-OFF: remove the chip by flowing dielectric (Hourmand et al., 2017).

cause the material to melt, vaporize, and then be removed from the workpiece after heating, as shown in Figure 9.23. Therefore, laser machining is a method of accurate material removal by heating the workpiece to a high temperature. The wavelength employed in the laser machining process determines how much energy is impinged on the material, which affects Material Removal Rate (MRR).

9.5.5.3 Plasma Arc Machining (PAM)

PAM, or Plasma-Arc Machining, is a technique for cutting metal using a tungsten inert-gas-arc torch that uses a high-velocity stream of high-temperature gases that melts and displaces materials in its path. The torch emits a high-velocity jet of plasma, a high-temperature ionized gas, which vaporizes and removes material from the workpiece to cut. In 1964, plasma welding technology became accessible to the welding industry as a way to improve control over the arc welding process at lower current levels. By offering a high degree of control and accuracy to create high-quality welds in small or precise tasks and by offering extended electrode life for large-scale manufacturing needs, plasma today still offers the advantages it first offered to the industry. When it was realized that an electric gas-cutting arc's performance could be substantially enhanced by being guided via a water-cooled copper

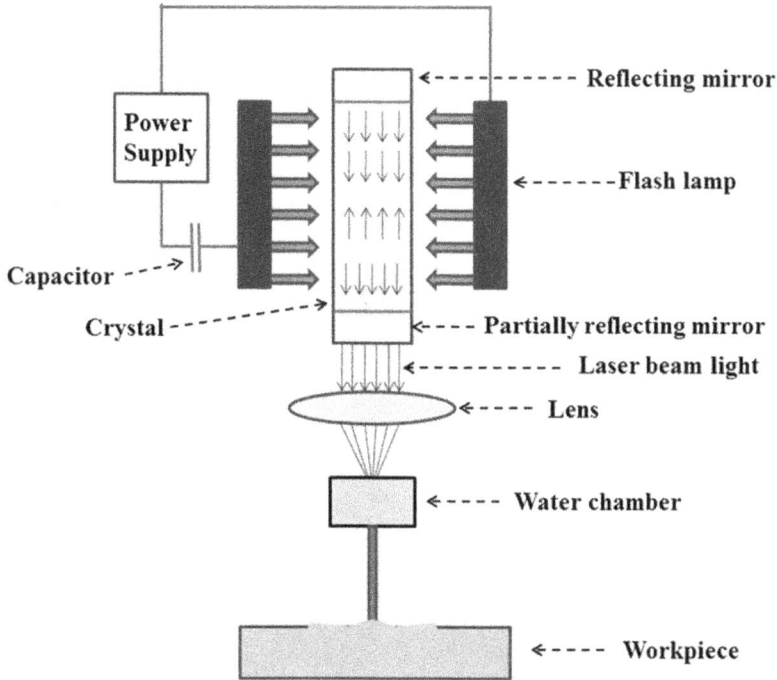

FIGURE 9.23 LBM schematic diagram (Kumar et al., 2021).

nozzle situated between an electrode (cathode) and the workpiece (anode), plasma arc cutting was invented in the early 1950s. Arc temperature and voltage rose as a result of the nozzle's ability to restrict the arc into a narrow cross-section. The arc was transformed into a fast-moving, extremely hot plasma jet after exiting the nozzle. This plasma jet powers the torch's cutting function. Plasma is a very hot, highly ionized gas that is created when a high-voltage electric arc passes through a mixture of gases comprising hydrogen, nitrogen, argon, and other gases. Plasma is frequently referred to as the fourth state of matter.

The heat source for the plasma arc-cutting process is a confined arc created by plasma gas. During this procedure, an electric arc is created between the workpiece and the electrode, with the electrode serving as the cathode. When electricity is fed through the plasma gas with the aid of a tungsten electrode, the plasma gas will expand through the nozzle at a high velocity, creating a high-intensity plasma arc. The surface being sliced receives this plasma arc, which converts some gas into plasma. This plasma arc is powerful enough to melt or vaporize the surface being cut, and it moves swiftly enough to allow molten metal to flow away from the cut area. Figure 9.24 depicts the PAM process. The majority of its applications are for cryogenic, high-temperature corrosion-resistant alloys. PAM is used to weld steel rocket motor boxes and nuclear submarine pipe systems.

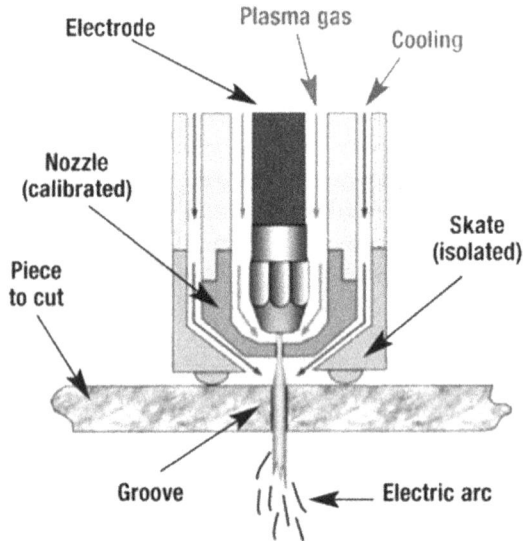

FIGURE 9.24   PAM process (Kunal Panchal, 2020).

9.6   CONCLUSION

This chapter proposes an overview of various machining processes and a few of the general comparison aspects between conventional and unconventional machining methods. The increasing demand for various alloys with upgraded physical and chemical properties has a larger impact on the development of the mechanical industry, where the conventional machining process is extensively used and showed promising chipping ability with traditional materials. However, the conventional machining process failed to realize its full potential for advanced materials and alloys. That is why there is a need for UCM, which leads to continuous research to bring forth the UCM process as a viable alternative to conventional machining. The UCM has the capability to shape complex shapes, has higher machinability, is computer-controlled, highly precise, and can micro-machine (miniaturization). Further, UCM capital and tooling costs, MRR, and high-power consumption are the major impediments that bind UCM to unfolding its full potential.

REFERENCES

Al-Ethari, H. A., Alsultani, K. F., & Dakhil, N. (2013). Variables affecting the chemical machining of stainless steel 420. *Stainless Steel*, *3*(6), 210–216.

Alajmi, M. S., & Almeshal, A. M. (2021). Modeling of cutting force in the turning of AISI 4340 using Gaussian process regression algorithm. *Applied Sciences*, *11*(9), 4055.

Allen, D. M., & Talib, T. N. (1983). Manufacture of stainless steel edge filters: An application of electrolytic photopolishing. *Precision Engineering*, *5*(2), 57–59.

Astakhov, V. P. (2011). Drilling. In *Modern Machining Technology* (pp. 79–212). Woodhead Publishing. https://doi.org/10.1533/9780857094940.79

Basu, B., Kalin, M., & Kumar, B. M. (2020). *Friction and Wear of Ceramics: Principles and Case Studies*. John Wiley & Sons.

Bhattacharyya, B., & Doloi, B. (2020). Machining processes utilizing chemical and electrochemical energy. In *Modern Machining Technology*. https://doi.org/10.1016/b978-0-12-812894-7.00005-0

Bhowmik, S., Jagadish, C., & Gupta, K. (2019). *Modeling and Optimization of Advanced Manufacturing Processes*. Springer.

Black, J. T., & Kohser, R. A. (2020). *DeGarmo's Materials and Processes in Manufacturing*. John Wiley & Sons.

Burger, M., Koll, L., Werner, E. A., & Platz, A. (2012). Electrochemical machining characteristics and resulting surface quality of the nickel-base single-crystalline material LEK94. *Journal of Manufacturing Processes*, *14*(1), 62–70. https://doi.org/10.1016/j.jmapro.2011.08.001

Çakir, O., Yardimeden, A., & Özben, T. (2007). Chemical machining. *Archives of Materials Science and Engineering*, *28*(8), 499–502.

Choudhary, S. K., & Jadoun, R. S. (2014). Latest research trend of optimization techniques in electric discharge machining (EDM). *Small*, *1*(02), 5mm.

Daud, N. D., Hasan, M. N., Saleh, T., Leow, P. L., & Mohamed Ali, M. S. (2022). Nontraditional machining techniques for silicon wafers. *The International Journal of Advanced Manufacturing Technology*, *121*(1–2), 29–57.

Deconinck, D., Van Damme, S., & Deconinck, J. (2012). A temperature dependent multi-ion model for time accurate numerical simulation of the electrochemical machining process. Part I: Theoretical basis. *Electrochimica Acta*, *60*, 321–328.

Dini, J. W. (1984). *Fundamentals of chemical milling*.

Drozda, T. J., & Wick, C. (1989). *Tool and Manufacuting Engineers Handbook (Chapter 14: Nontraditional Machining)*. SME Publisging.

El-Hofy, H. (n.d.). *Advanced Machining Processes: Nontraditional and Hybrid Machining Processes*. *2005*. McGraw Hill Professional.

Girsang, I. P., & Dhupia, J. S. (2015). Machine tools for machining. In *Handbook of Manufacturing Engineering and Technology* (pp. 811–865). Springer London. https://doi.org/10.1007/978-1-4471-4670-4_4

Haider, J., & Hashmi, M. S. J. (2014). 8.02: Health and environmental impacts in metal machining processes. *Comprehensive Materials Processing*, *8*, 7–33.

Harris, W. T. (1976). *Chemical Milling: The Technology of Cutting Materials by Etching*. Clarendon Press.

Hourmand, M., Sarhan, A. A. D., Noordin, M. Y., & Sayuti, M. (2017). 1.10 micro-EDM drilling of tungsten carbide using microelectrode with high aspect ratio to improve MRR, EWR, and hole quality. In *Comprehensive Materials Finishing* (pp. 267–321). Elsevier. https://doi.org/10.1016/B978-0-12-803581-8.09155-4

Joshi, A. Y., & Joshi, A. Y. (2019). A systematic review on powder mixed electrical discharge machining. *Heliyon*, *5*(12), e02963. https://doi.org/10.1016/j.heliyon.2019.e02963

Kalman, H., Tripathi, N. M., Gabrieli, O. G., & Portnikov, D. (2017). Phase diagrams for pneumatic and hydraulic conveying. In *18th International Conferences on Transport and Sedimentation of Solid Particles, T and S 2017* (pp. 145–152). Wydawnictwo Uniwersytetu Przyrodniczego we Wroclawiu.

Kumar, S., Roy, D. N., & Dey, V. (2021). A comprehensive review on techniques to create the anti-microbial surface of biomaterials to intervene in biofouling. *Colloid and Interface Science Communications*, *43*, 100464. https://doi.org/10.1016/j.colcom.2021.100464

Lauwers, B., Kruth, J.-P., Liu, W., Eeraerts, W., Schacht, B., & Bleys, P. (2004). Investigation of material removal mechanisms in EDM of composite ceramic materials. *Journal of Materials Processing Technology*, *149*(1–3), 347–352.

Lumay, G., Tripathi, N. M., & Francqui, F. (2019). How to gain a full understanding of powder flow properties, and the benefits of doing so. *ONdrugDelivery*, *2019*, 42–47.

McGeough, J. A. (1974). *Principles of Electrochemical Machining*. CRC Press.

M Tripathi, N., & S Mallick, S. (2017). Pneumatic conveying of Fly Ash: Bend Models investigation. *Advanced Materials Proceedings*, *2*(8), 526–531.

Neveu, A., Lumay, G., Pillitteri, S., Monsuur, F., Pauly, T., Ribeyre, Q., Francqui, F., Vandewalle, N., & Tripathi, N. M. (2020b). Physical characterization of blends containing mesoporous particles with a focus on electrostatic properties. In *2020 AIChE Spring Meeting and 16th Global Congress on Process Safety*. AIChE.

Neveu, A., Tripathi, N. M., Rigo, O., Francqui, F., & Lumay, G. (2020a). Experimental investigation of spreadability of metal powders in recoating process. In *Proceedings - Euro PM2020 Congress and Exhibition (Proceedings - Euro PM2020 Congress and Exhibition)*. European Powder Metallurgy Association (EPMA).

Nguyen-Dinh, N., Hejjaji, A., Zitoune, R., Bouvet, C., & Crouzeix, L. (2018). Machining of FRP composites: Surface quality, damage, and material integrity: Critical review and analysis. *Futuristic Composites: Behavior, Characterization, and Manufacturing*, 1–35.

Palanikumar, K., & Davim, J. P. (2013). Electrical discharge machining: Study on machining characteristics of WC/Co composites. In *Machining and Machine-Tools* (pp. 135–168). Woodhead Publishing.

Panchal, K. S., & Mungla, M. (2020). A review on optimization of plasma arc cutting parameters using Taguchi method for EN19. *Materials Science*, *10*, 172–191. https://doi.org/10.46243/jst.2020.v5.i3.pp172-191

Pandilov, Z. (2018, March). Application of electro chemical machining for materials used in extreme conditions. In *IOP Conference Series: Materials Science and Engineering* (Vol. 329, No. 1, p. 012014). IOP Publishing.

Pedroso, A. F. V., Sousa, V. F. C., Sebbe, N. P. V., Silva, F. J. G., Campilho, R. D. S. G., Sales-Contini, R. C. M., & Jesus, A. M. P. (2023). A comprehensive review on the conventional and non-conventional machining and tool-wear mechanisms of INCONEL®. *Metals*, *13*(3), 585.

Petrofes, N. F., & Gadalla, A. M. (1988). Processing aspects of shaping advanced materials by electrical discharge machining. *Material and Manufacturing Process*, *3*(1), 127–153.

Prasad, K., & Chakraborty, S. (2018). A decision guidance framework for non-traditional machining processes selection. *Ain Shams Engineering Journal*, *9*(2), 203–214.

Raj, A., Anirudh, V., & Nanjundeswaraswamy, T. S. (2019). Chemical machining process-an overview. *International Research Journal of Innovations in Engineering and Technology*, *3*(11), 37.

Ratsimba, A., Zerrouki, A., Tessier-Doyen, N., Nait-Ali, B., André, D., Duport, P., Neveu, A., Tripathi, N., Francqui, F., & Delaizir, G. (2021). Densification behaviour and three-dimensional printing of Y2O3 ceramic powder by selective laser sintering. *Ceramics International*, *47*(6), 7465–7474.

Rubio, E., & Jáuregui-Correa, J. C. (2012). A wavelet approach to estimate the quality of ground parts. *Journal of Applied Research and Technology*, *10*(1 SE-Articles). https://doi.org/10.22201/icat.16656423.2012.10.1.418

Sanchez, J. A., Cabanes, I., Lopez de Lacalle, L. N., & Lamikiz, A. (2001). Development of optimum electrodischarge machining technology for advanced ceramics. *The International Journal of Advanced Manufacturing Technology*, *18*, 897–905.

Sharma, A., Babbar, A., Tian, Y., Pathri, B.P., Gupta, M., & Singh, R. (2022). Machining of ceramic materials: A state of the art review. *International Journal on Interactive Design and Manufacturing (IJIDeM)*, 1–21. https://doi.org/10.1007/s12008-022-01016-7

Sharma, A., & Jain, V. (2020). Experimental investigation of cutting temperature during drilling of float glass specimen. In *IOP Conference Series: Materials Science and Engineering* (Vol. 715, No. 1, p. 012050). IOP Publishing.

Sharma, A., Jain, V., & Gupta, D. (2018). Characterization of chipping and tool wear during drilling of float glass using rotary ultrasonic machining. *Measurement*, *128*, 254–263.

Sharma, A., Jain, V., & Gupta, D. (2019a). Tool wear analysis while creating blind holes on float glass using conventional drilling: A multi-shaped tools study. In *Advances in Manufacturing Processes: Select Proceedings of ICEMMM 2018* (pp. 175–183). Springer Singapore.

Sharma, A., Jain, V., & Gupta, D. (2019b). Comparative analysis of chipping mechanics of float glass during rotary ultrasonic drilling and conventional drilling: For multi-shaped tools. *Machining Science and Technology*, *23*(4), 547–568.

Sharma, A., Jain, V., & Gupta, D. (2019c). Multi-shaped tool wear study during rotary ultrasonic drilling and conventional drilling for amorphous solid. *Proceedings of the Institution of Mechanical Engineers, Part E: Journal of Process Mechanical Engineering*, *233*(3), 551–560.

Sharma, A., Jain, V., & Gupta, D. (2021a). Effect of pre and post tempering on hole quality of float glass specimen: For rotary ultrasonic and conventional drilling. *Silicon*, *13*, 2029–2039.

Sharma, A., Jain, V., & Gupta, D. (2021b). Mathematical approach on chipping volume estimation generated during rotary ultrasonic drilling for float glass. *Proceedings of the National Academy of Sciences, India Section A: Physical Sciences*, *92*, 285–291.

Sharma, A., Jain, V., Gupta, D., & Babbar, A. (2020). A review study on miniaturization. In *Advanced Manufacturing and Processing Technology* (First edition, pp. 111–131). CRC Press.

Sharma, A., Kalsia, M., Uppal, A. S., Babbar, A., & Dhawan, V. (2022). Machining of hard and brittle materials: A comprehensive review. *Materials Today: Proceedings*, *50*, 1048–1052.

Singh, K. J., Ahuja, I. S., & Kapoor, J. (2018). Ultrasonic, chemical-assisted ultrasonic and rotary ultrasonic machining of glass: A review paper. *World Journal of Engineering*, *15*(6), 751–770. https://doi.org/10.1108/WJE-04-2018-0114

Srikanth, D. V, & Rao, M. S. (2014). Application of optimization methods on abrasive jet machining of ceramics. *International Journal of Industrial Engineering & Technology (IJIET)*, *4*(3), 23–32.

Taniguchi, N. (1983). Current status in, and future trends of, ultraprecision machining and ultrafine materials processing. *CIRP Annals*, *32*(2), 573–582.

Tehrani, A. F., & Imanian, E. (2004). A new etchant for the chemical machining of St304. *Journal of Materials Processing Technology*, *149*(1–3), 404–408.

Toenshoff, H. K., & Denkena, B. (2013). *Basics of Cutting and Abrasive processes*. Springer.

Tripathi, N. M., Francqui, F., & Lumay, G. (2020). Influence of relative air humidity on the flow property of fine powders. In *Third International Conference on Powder, Granule and Bulk Solids: Innovations and Applications PGBSIA 2020 February 26–28, 2020* (p. 63).

Tripathi, N. M., Francqui, F., Pirenne, T., & Lumay, G. (2019). Measuring food powders electrical properties as a result of anti-static content. In *2019 AIChE Annual Meeting*. American Institute of Chemical Engineers.

Tripathi, N. M., Levy, A., & Kalman, H. (2016). Initial acceleration pressure drop in dilute phase pneumatic conveying system. In *Powder, Granule and Bulk Solids: Innovations and Applications Conference*.

Tripathi, N. M., & Mallick, S. S. (2014). *An Investigation into Pressure Drop Across Bends for Fluidised Densephase Pneumatic Conveying Systems* (Doctoral dissertation).

10 Future Trends and Challenges in Super-finishing Processes
Post-Metal Additive Manufacturing Process

Neetesh Soni, Gilda Renna, and Paola Leo

10.1 INTRODUCTION

Metal parts fabrication with Additive Manufacturing (AM) has become increasingly popular in various fields due to its design flexibility and ability to produce parts with complex geometries. Among the various sectors, the aerospace and biomedical sectors are the most interested in the development of this technique, as high-performance materials such as titanium alloys, stainless steels, Inconel, etc. are used. However, the surface finish of AM products very often does not meet the required specifications, thus making post-processing of the AM part essential to achieve a better surface finish and/or desired mechanical property. Therefore, today several research studies are aimed at the development of post-processing techniques that are efficient and effective to improve the quality of the surface properties and therefore the functionality of metal AM products. The selection of super-finishing techniques is influenced by critical factors such as the microstructural characteristics and mechanical properties of metallic AM products. In detail the microstructure of a metallic AM product refers to its internal structure at the microscopic level, including the arrangement of grains, the presence of defects, and phases within the material. The unique microstructural characteristics of AM products, such as grain size, grain orientation, and the presence of voids or porosity, can influence the choice of super-finishing technique. For example:

1) **Grain Size**: AM processes can produce fine-grained or coarse-grained structures, depending on the process parameters and heat treatment. Finer-grained materials generally exhibit improved mechanical properties and surface finish. In such cases, super-finishing methods like abrasive polishing or buffing may be sufficient to achieve the desired surface quality.

DOI: 10.1201/9781003428862-10

Coarser-grained materials, on the other hand, may require more aggressive super-finishing techniques, such as honing or lapping, to refine the surface.

2) **Porosity and Voids**: AM products may contain inherent porosity or voids due to the layer-by-layer deposition process. These voids can impact the surface finish and mechanical properties of the material. Super-finishing methods that can effectively reduce or fill in these voids, such as vibratory finishing or impregnation, may be necessary to achieve a smooth and defect-free surface.

10.1.1 MECHANICAL PROPERTIES

The mechanical properties of metallic AM products, including hardness, tensile strength, and ductility, can also influence the selection of super-finishing techniques. Different super-finishing methods have varying effects on the mechanical properties of the material. Considerations include:

1) **Hardness**: AM materials can exhibit a wide range of hardness levels depending on factors such as alloy composition and heat treatment. Super-finishing techniques must be chosen carefully to avoid excessive material removal or surface damage in harder materials. Methods such as chemical or electrochemical polishing, which operate at lower material removal rates, may be preferred for hard materials to achieve a fine surface finish while preserving the mechanical properties.

2) **Ductility**: Some AM processes can introduce residual stresses in the material, affecting its ductility. Super-finishing techniques that involve high-pressure or severe plastic deformation, such as roller burnishing or shot peening, can potentially relieve residual stresses, and improve the ductility of the material while enhancing the surface finish.

In this chapter, the main super-finishing techniques that can be employed to enhance the surface finish of AM products will be highlighted. These include abrasive machining (Deja et al.,2021), sanding, mass finishing (Khorasani et al., 2022), coating (Pathak et al., 2017), burnishing (Salmi et al., 2017), bonding, and electrochemical polishing. Each of these techniques has its advantages and limitations, and the selection of the appropriate technique depends on the specific requirements of the product and applications. The challenges faced by manufacturers in achieving the desired surface finish, such as the complex geometries (less than 1 mm radius holes) of AM products, the residual stresses and defects in the products, and the high surface roughness of AM products will also be discussed . In order to overcome these challenges, possible solutions such as process optimization, the use of specialized tools, and the integration of multiple techniques will be suggested. Few studies are also moving towards the future trends in finishing, machining, and additive manufacturing processes: post-metal additive manufacturing processes (Sharma et al., 2022; Sharma & Jain, 2020; Sharma et al., 2018; Sharma et al., 2019a; Sharma et al., 2019b; Sharma et al., 2019c; Sharma et al., 2021a; Sharma et al., 2021b; Sharma et al., 2020; Sharma et al., 2022; Kalman et al., 2017; Lumay et al., 2019; Tripathi

& Mallick, 2017; Neveu et al., 2020a; Neveu et al., 2020b; Ratsimba et al., 2021; Tripathi et al., 2020; Tripathi et al., 2019; Tripathi et al., 2016; Tripathi and Mallick, 2014).

Finally, the chapter proposes several future research directions to enhance the effectiveness of super-finishing techniques after the completion of the metal AM process. These include the development of new super-finishing techniques specifically tailored for AM products, the optimization of existing techniques for AM products, and the exploration of the use of artificial intelligence and machine learning in super-finishing processes. Overall, the literature highlights the importance of super-finishing processes in enhancing the quality and functionality of metal AM products and emphasizes the need for continued research in this field to further improve the effectiveness of these processes.

10.1.2   ADDITIVE MANUFACTURING: MATERIALS, PROCESS, CLASSIFICATION

Additive Manufacturing, also known as 3D printing, is a process of building objects by adding layers of material one at a time. The materials used in AM can vary depending on the application and the specific process used. Some common materials used in AM include metals and alloys which are our targeted materials. The process of AM involves three main steps: design, slicing, and printing.

The design step involves creating a digital 3D model of the object to be printed using Computer-Aided Design (CAD) software. The slicing step involves dividing the digital model into thin horizontal layers, which serve as a guide for the printing process. The printing step involves using a printer to add layers of material on top of each other until the object is complete (Mohanavel et al., 2021).

AM can be classified into several categories based on the specific process used. Some common AM processes include Selective Laser Melting (SLM) (Salmi 2021), Fused Deposition Modeling (FDM), Stereolithography (SLA), and Binder Jetting (BJ). Each process has its unique advantages and limitations, and the choice of process depends on the desired outcome and the properties of the materials being used. SLM, for example, is a process that involves melting a powder and depositing it layer by layer to create an object. SLM, on the other hand, uses a laser to solidify in a dwell time into the desired shape. SLM involves using a laser in a vacuum environment which may contain nitrogen, carbon dioxide or any one of them. To protect against high-temperature oxidation during the manufacturing of objects, the object's manufacturing and surface finishing are correlated due to pre- and post-processing factors that affect the surface finish. Several factors affect the surface finish, such as process parameters, material selection, part-building orientation, layer strategy, and surface post-processing. Figure 10.1 provides a detailed explanation of the complete factors involved in the surface finishing process, including machine parameters and post-processing steps (Peng et al., 2021).

AM is a versatile technology as it allows not only the creation of parts in metallic material but also in other materials such as plastics, composites, and ceramics. During the process, there are parameters of paramount importance that need to be considered such as layer thickness, scan distance, scan angle, laser source, hatch

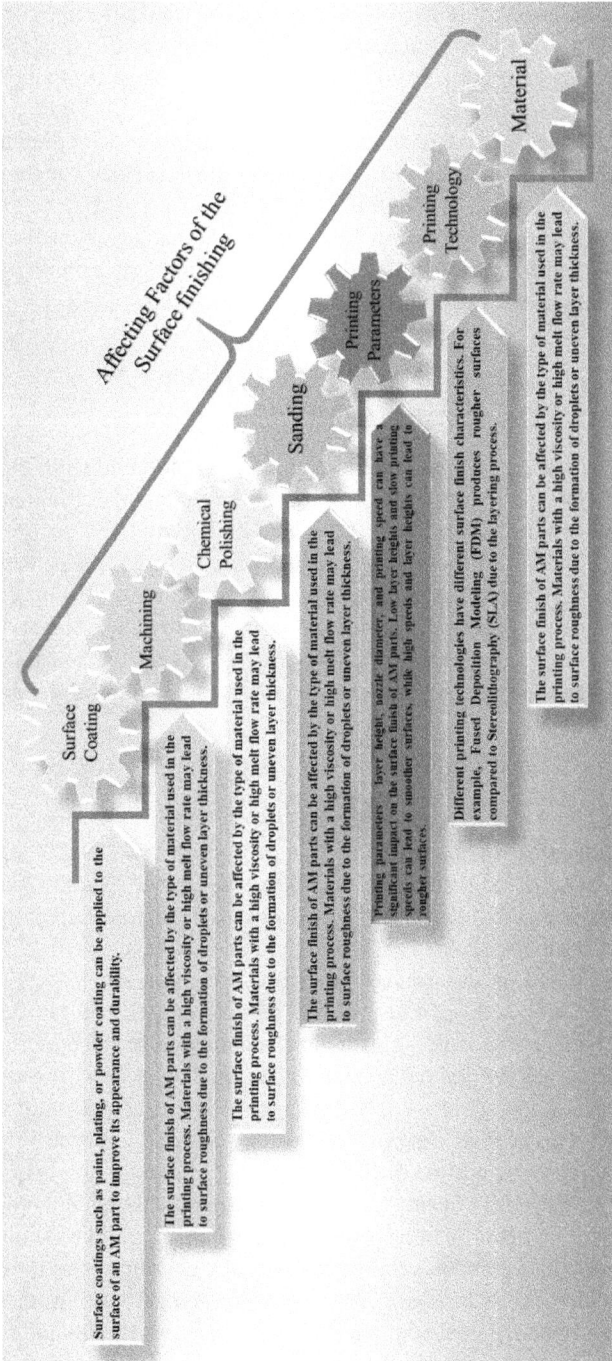

FIGURE 10.1 Factors affecting the surface finishing and demands increasing for the smooth products.

distance, and other process parameters. The classification of AM processes provides a framework for understanding the different approaches and selecting the appropriate technique for a given application. The continuous development of new materials and processes in AM is likely to further expand its range of applications and capabilities.

The Additive Manufacturing (AM) super-finishing processes are used to improve the surface quality of 3D printed parts by removing any surface irregularities and improving the surface finish. In Figure 10.1 already mentioned some factors that can affect the super-finishing process in additive manufacturing such as material selection, part orientation, layer thickness, printing technology, finishing technique, surface preparation and equipment etc.

10.2 ADDITIVE MANUFACTURING APPLIED POST-PROCESSING ON PARTS/OBJECTS

The application of the post-processing operations on AM-manufactured parts basically depends on the technology used. For example, some AM techniques may not require further machining as they offer good surface quality.

AM processes typical of metallic materials (such as SLM, PBSL, sintering, or others) always require further post-production treatments such as support removal, sanding, mass finishing, vapour smoothing, painting, electro-polishing, bonding etc. (Seifi et al., 2017). Moreover, the choice of post-processing method to be applied depends not only on the type of AM technology but also on the raw material used.

10.2.1 SUPPORT REMOVAL

Support removal is a crucial post-processing operation in AM that involves the removal of support structures used to hold up a part during the printing process. Typically, these support structures are added to prevent deformation, warping, or collapse of the printed part during the printing process. However, once the printing is complete, these must be removed to achieve the final desired shape and finish of the part. Support removal is used in various industries, including aerospace, automotive, medical, and consumer goods. In aerospace and automotive, AM is often used to produce complex geometries and lightweight structures, making support removal a critical step in achieving the desired product. In the medical field, AM is used to produce custom implants and prosthetics, where support removal is necessary to achieve a precise fit and surface finish. Support removal is also used in consumer goods for the production of high-end custom products, such as jewellery and artwork. Support removal in AM has several advantages. First, it allows for the creation of complex geometries that are difficult or impossible to achieve using traditional manufacturing techniques. Second, it enables the production of lightweight and durable parts, making it suitable for the aerospace and automotive industries. Third, it offers a cost-effective and time-efficient solution for small-scale production and customization. Moreover, support removal in AM has some limitations as well. The removal process can be time-consuming and require skilled labour. The support structures can also leave behind marks or blemishes on the surface of the part, affecting the final surface finish. Additionally, removing support structures can damage the part, requiring additional post-processing steps. As the technology continues to evolve, new solutions for support removal may arise, leading to faster and more efficient post-processing techniques (Dizon et al., 2021).

10.2.2 SANDING

Sanding is a surface finishing process used to remove surface irregularities, such as roughness, scratches, and tool marks, from a material's surface. Sanding involves the use of abrasive materials, such as sandpaper or abrasive pads, to smooth and refine the surface of the material.

It is used in a wide range of industries, including woodworking, metalworking, automotive, and construction, for example.

1) In woodworking, sanding is used to smooth and prepare surfaces for painting or staining.
2) In metalworking, sanding is used to remove burrs and prepare surfaces for welding or painting.
3) In the automotive industry, sanding is used to prepare surfaces for painting or polishing.
4) In construction, sanding is used to prepare surfaces for plastering or painting.

Sanding has several advantages. First, it is a cost-effective method of achieving a smooth surface finish. Second, it is a relatively simple process that requires minimal training or equipment. Third, it is a flexible process that can be used on a variety of materials, including wood, metal, and plastic. Moreover, sanding has some limitations. First, it is a time-consuming process, particularly for large surfaces. Second, sanding can create dust, which can be hazardous to health if not handled properly. Third, sanding can remove too much material, leading to dimensional inaccuracies and reduced structural integrity. Sanding is a widely used surface finishing process that is suitable for a variety of materials and applications. As technology continues to advance, new sanding techniques may emerge, making the process faster, more efficient, and more precise.

10.2.3 MASS FINISHING

Mass finishing is another surface finishing process that involves the use of abrasive media, chemicals, or other finishing agents to achieve a desired surface finish on a large number of parts at the same time. The process is typically carried out in a vibratory tumbler or a rotary barrel, where the parts and finishing media are mixed and agitated. Mass finishing is used in a variety of industries, including automotive, aerospace, jewellery, and metalworking. In automotive and aerospace, mass finishing is used to deburr, polish, and clean metal components, such as engine parts, gears, and turbine blades. In jewellery making, mass finishing is used to achieve a uniform surface finish on metal parts, such as clasps and beads. In metalworking, mass finishing is used to remove surface defects, such as casting lines and welding marks, from metal parts.

It has several advantages. First, it is a cost-effective process that can produce a uniform finish on a large number of parts at the same time. Second, it is a versatile process that can be used on a wide range of materials, including metals, plastics, and ceramics. Third, it is a relatively simple process that requires minimal training or equipment. Moreover, mass finishing has some limitations. First, it can be time-consuming, particularly for large parts or batches of parts. Second, it may not be suitable for parts with complex geometries or tight tolerances, as the finishing media may not reach all surfaces equally. Third, it can be noisy and may require soundproofing to protect workers' hearing. Mass finishing is a widely used surface finishing process that is suitable for a variety of industries and applications.

10.2.4 Painting or Coating

Painting or coating is a surface finishing process that involves applying a layer of paint or coating material to a substrate to improve its appearance, protect it from corrosion or wear, or add functionality. The process involves a series of steps, including surface preparation, application of the paint or coating material, and drying or curing. Painting or coating is widely used in many industries, painting is used to protect the vehicle from corrosion and to improve its appearance. In the aerospace industry, coatings are used to protect aircraft from high temperatures, corrosion, and wear. In construction, painting is used to protect buildings from weathering and to add aesthetic appeal (Methew et al., 2023 and Gray & Luan 2002). It has numerous advantages. First, it can improve the appearance of the substrate, making it more attractive and appealing. Second, it can protect the substrate from environmental factors, such as corrosion, UV radiation, and wear. Third, it can add functionality, such as anti-graffiti coatings, anti-slip coatings, and fire-resistant coatings. Painting or coating has some limitations. The process can be time-consuming, requiring multiple coats and drying or curing time. Second, the process can be costly, particularly for large-scale applications or when high-quality coatings are required. Third, coatings may not adhere well to certain substrates or may peel off over time. As technology continues to advance, new coatings and painting techniques may emerge, making the process more efficient, cost-effective, and environmentally friendly.

10.2.5 Electroplating

Electroplating involves the deposition of a thin layer of metal onto a substrate using an electric current. The process involves immersing the substrate in a solution containing metal ions and passing an electric current through the solution to attract the metal ions onto the substrate. In the automotive industry, electroplating is used to improve the appearance and durability of parts, such as chrome plating on bumpers and wheels. In the aerospace industry, electroplating is used to improve the corrosion resistance and wear properties of parts. In electronics, it is used to create conductive traces on printed circuit boards. In jewellery, it is used to create a thin layer of gold or other precious metals onto a base metal. Electroplating has several merits. First, it can improve the appearance of the substrate, making it more attractive and appealing. Second, it can increase the durability and wear resistance of the substrate. Third, it can improve the substrate's electrical conductivity and corrosion resistance. It has some limitations, that is, it can be a complex and time-consuming process, requiring careful preparation and monitoring. Also, it can be expensive, particularly for large-scale applications or when high-quality plating is required, and it can be environmentally hazardous due to the use of toxic chemicals.

10.2.6 BONDING

Bonding surface finishing methods are techniques used to modify the surface of a material to improve its adhesion to another material during bonding. Surface finishing methods can include physical, chemical, or mechanical treatments, and they are essential to ensuring a strong bond between two materials. Bonding surface finishing is commonly used in industries such as electronics, automotive, and aerospace. Bonding surface finishing methods are used in a wide range of applications, including:

1) **Electronics**: In the electronics industry, bonding surface finishing is used to ensure strong adhesion between printed circuit boards and electronic components.
2) **Automotive**: Bonding surface finishing is used in the automotive industry to improve the bond strength between various parts of a vehicle.
3) **Aerospace**: The aerospace industry relies heavily on bonding surface finishing to ensure the integrity and reliability of components in airplanes and spacecraft (Duan et al., 2021).
4) **Medical Devices**: Bonding surface finishing is used in the medical device industry to ensure the adhesion of components such as sensors, electrodes, and implants (Bowman and Meindl, 1986).

The use of bonding surface finishing methods offers several merits, including:

1) **Improved Adhesion**: The main advantage of bonding surface finishing is that it improves the adhesion between materials, resulting in a stronger bond.
2) **Increased Durability**: Bonding surface finishing can also improve the durability of a bond by providing a protective layer to the surface.
3) **Versatility**: Bonding surface finishing methods are highly versatile and can be used on a wide range of materials.

Despite its advantages, bonding surface finishing also has some limitations, including:

1) **Surface Preparation**: Bonding surface finishing methods require careful surface preparation, which can be time-consuming and costly.
2) **Environmental Concerns**: Some surface finishing methods can be harmful to the environment, so special care must be taken in their use and disposal. Overall, bonding surface finishing is an important technique that enables strong adhesion between materials and finds widespread application in numerous industries.

10.2.7 LASER POLISHING

Laser polishing is a surface finishing process that uses a high-powered laser beam to melt and smooth out the surface of a material. It is a non-contact method that can be used to achieve high precision and control in surface finishing. Laser polishing is used in a variety of industries, including aerospace, automotive, medical, and electronics. It can be used to finish metal parts, ceramics, and plastics, and is particularly useful for parts that have complex geometries or hard-to-reach areas. One of the main advantages of laser polishing is its precision and control. The laser beam can be focused very precisely, allowing for very precise removal of material (Krishnan & Fang, 2019). Laser polishing can also be used to achieve a very smooth surface finish, with a roughness of fewer than 0.1 micrometres in some cases. Additionally, laser polishing is a non-contact process, which means that it can be used to finish delicate parts without damaging them. One limitation of laser polishing is that it can only be used on materials that can absorb the laser beam. This means that it is not suitable for all materials, particularly those that are transparent or reflective. Additionally, laser polishing can be a slow process, particularly for larger parts, which can limit its usefulness in some applications. Its limitations mean that it may not be suitable for all materials or applications, and other surface finishing techniques may be more appropriate in some cases.

10.2.8 ULTRASONIC ABRASIVE POLISHING

Ultrasonic abrasive polishing uses ultrasonic vibrations to apply abrasive particles to a surface, resulting in the removal of material and the creation of a polished surface. This process is typically used for finishing hard and brittle materials, such as ceramics, glass, and certain metals. Ultrasonic abrasive polishing is commonly used in the manufacturing of precision components, such as optical lenses, semiconductor wafers, and medical implants. It can also be used for the restoration of antique or damaged items, such as glassware and ceramics. One of the primary advantages of ultrasonic abrasive polishing is its ability to produce a highly uniform and precise surface finish. The use of ultrasonic vibrations helps to evenly distribute abrasive particles across the surface, resulting in a consistent removal of material. This process also allows for very fine finishes, with roughness values as low as 0.01 micrometres achievable. Additionally, ultrasonic abrasive polishing is a relatively gentle process that does not produce significant heat or mechanical stress, which can be beneficial for delicate materials (Jones & Hull, 1998). One limitation of ultrasonic abrasive polishing is that it can be a time-consuming process, particularly for larger parts or complex geometries. Additionally, the process can be expensive due to the cost of specialized equipment and abrasive materials. Finally, ultrasonic abrasive polishing may not be suitable for all materials, as some materials may not respond well to the abrasive particles or ultrasonic vibrations. It offers many advantages over other surface finishing techniques and is well-suited for finishing hard and brittle materials.

10.2.9   Surface Roughness Measuring Technique

Surface roughness refers to the irregularities and deviations present on the surface of a material or component. It is an important characteristic that can affect the performance and quality of a product. Surface roughness can be quantified by measuring the average deviation of surface heights from a reference plane. The measurement of surface roughness is necessary for several reasons. First, it can affect the performance of a product. For example, if a component has a rough surface, it may cause more friction and wear, leading to reduced efficiency and a shorter lifespan. Second, it can affect the appearance of a product. If the surface of a product is rough or uneven, it may be perceived as lower quality or less attractive. Moreover, surface roughness can affect the ability of two surfaces to adhere to each other. If the surface is too rough, it may prevent proper bonding or adhesion.

Measuring surface roughness has some advantages. It allows for quality control of products and can help identify defects or inconsistencies in manufacturing processes. It can also aid in the selection of appropriate materials and surface treatments for specific applications. There are also some limitations and drawbacks to measuring surface roughness. The accuracy of measurements can be affected by factors such as the material being measured, the measurement technique used, and the surface finish of the reference standard. Surface roughness measurement is used in a variety of industries, including manufacturing, automotive, aerospace, and medical devices (Horn & Harrysson, 2012). It is used to ensure that products meet specific quality standards and to identify areas for improvement in manufacturing processes. For example, in the automotive industry, surface roughness measurement is used to ensure that engine components meet specific tolerances for friction and wear. In the medical device industry, surface roughness measurement is used to ensure that implants and other medical devices are compatible with the body and do not cause excessive wear or damage. Surface roughness is an important consideration in additive manufacturing, where the surface finish of a printed part can affect its performance and quality. In some cases, the roughness of a printed part may be desirable, such as in applications where adhesion or friction is required. In other cases, a smoother surface finish may be necessary for proper function or appearance. Manufacturers can use surface roughness measurement to ensure that printed parts meet specific tolerances for surface finish and to identify areas for improvement in the additive manufacturing process. There are several techniques for measuring surface roughness, each with its own advantages and limitations. Here are some of the most commonly used techniques such as contact profilometry, optical interferometry, Atomic Force Microscopy (AFM), Scanning Electron Microscopy (SEM), laser scanning confocal microscopy, white light interferometry, stylus profilometry, and optical scatterometry. These are explained in detail in Table 10.1

10.3 FUTURE APPLICATION BY AM

Even today this technology continues to evolve and is expected to have a significant impact on future manufacturing applications. Additive manufacturing has the potential to revolutionize manufacturing by enabling the production of parts and products that were previously impossible to produce. Some of the future applications of AM could include:

1) **Bio-printing**: AM can be used to produce tissues and organs using living cells. Bio-printing has the potential to revolutionize medicine by providing personalized and precise treatments.
2) **Aerospace**: AM can enable the production of lightweight, complex parts that are difficult to produce using traditional manufacturing methods. This can result in significant weight savings and increased efficiency in aircraft.
3) **Construction**: AM can be used to produce complex building components, such as walls and facades, with reduced waste and increased design freedom (Buchanan, 2019).
4) **Energy**: AM can enable the production of complex energy-efficient parts and components, such as turbines and heat exchangers, that can improve the performance and efficiency of energy systems.

The future applications of AM offer several advantages, including:

1) **Design Freedom**: AM enables the production of parts and products with complex geometries and customized features, providing designers with greater design freedom.
2) **Reduced Waste**: AM can reduce material waste by producing parts with optimized geometries that require less material (Jiang et al., 2018).
3) **Faster Production**: AM can produce parts and products in a shorter time frame than traditional manufacturing methods, resulting in faster production times.

Despite its many advantages, AM also has some limitations, including AM can be expensive compared to traditional manufacturing methods, particularly for larger parts and products. The range of materials that can be used in AM is still somewhat limited, although this is expected to improve as the technology evolves (Mohsen, 2017). AM has the potential to transform manufacturing and enable the production of parts and products with unprecedented design flexibility and customization. Its future applications are expected to impact a wide range of industries and offer numerous benefits, although some limitations will need to be addressed as the technology continues to evolve.

TABLE 10.1

Surface Roughness Measuring Methods and Techniques in Details

Techniques	Details	Ref.
Contact profilometry	This method involves physically touching the surface with a stylus that moves across the surface and measures the vertical displacement of the stylus. The resulting data are used to generate a 3D map of the surface roughness. This method is highly accurate and versatile, but it can only be used on relatively flat surfaces.	(Kim & Park, 2019)
Optical interferometry	This method uses a light source to project a pattern onto the surface being measured and then measures the distortion of the pattern caused by the surface roughness. This technique can be used to measure very small surface features, but it is less accurate than contact profilometry.	(Krishnan & Fang, 2019)
Atomic Force Microscopy (AFM	This method uses a sharp probe to scan the surface of the sample and measures the forces between the probe and the surface. This technique can be used to measure surface roughness at the nanometer scale, but it is relatively slow and expensive.	(Kim & Park, 2019)
Scanning Electron Microscopy (SEM)	This method uses a beam of electrons to scan the surface of the sample and measures the electrons that are scattered back. This technique can be used to generate highly detailed images of the surface, but it is limited in its ability to measure surface roughness directly.	(Kim & Park, 2019)
Laser scanning confocal microscopy	This method uses a laser to scan the surface of the sample, and measures the reflected light. By analyzing the intensity of the reflected light, a 3D map of the surface roughness can be generated. This method is highly accurate and can be used on a variety of surfaces, but it is relatively expensive.	(Ma et al., 2017)
White Light Interferometry	This method uses a low-coherence white light source and an interferometer to measure the surface topography with nanometre resolution. The light is focused on the surface, and the reflected light is detected by the interferometer. The interference pattern generated by the reflected light is analyzed to determine the surface roughness.	(Carmignato et al., 2020)
Stylus profilometry	This method is similar to contact profilometry, but it uses a diamond-tipped stylus to scan the surface. The vertical displacement of the stylus is measured as it moves across the surface, and these data are used to generate a 3D map of the surface roughness. This method is relatively inexpensive and can be used on a variety of surfaces, but it is limited to measuring surfaces that are relatively flat.	(Ma et al., 2017)
Optical scatterometry	This method uses light scattering to measure surface roughness. A beam of light is directed at the surface and the scattered light is detected and analyzed to determine the surface roughness. This method is noncontact and can be used to measure surfaces with complex shapes, but it is less accurate than some of the other methods.	(Carmignato et al., 2020)

10.4 FUTURISTIC APPROACH

In addition to the current post-processing techniques mentioned, there are several futuristic approaches being explored to further improve the surface quality of Additively Manufactured (AM) parts. These approaches aim to address the challenges associated with poor surface quality and inadequate mechanical properties in metal AM. Here are some details about these futuristic approaches:

1) **Topological Optimization**: Topological optimization is a design approach that optimizes the material distribution within a given design space to achieve the desired mechanical performance. By using this approach in conjunction with AM, it is possible to create designs that inherently have better surface quality and mechanical properties, reducing the need for extensive post-processing.

2) **In-process Monitoring and Control**: Real-time monitoring and control systems integrated into the AM process can help identify and address issues that contribute to poor surface quality and mechanical properties. These systems can monitor parameters such as temperature, energy input, melt pool characteristics, and part distortion. By actively controlling these parameters during the printing process, the quality of the final part can be improved.

3) **Hybrid Manufacturing**: Hybrid manufacturing combines additive manufacturing with traditional subtractive processes, such as milling or grinding, to achieve higher precision and surface finish. By using AM to build the near-net shape and then employing subtractive processes for final machining, the surface quality can be significantly improved.

4) **Advanced Material Development**: Researchers are actively developing new materials specifically tailored for AM processes. These materials have improved microstructural characteristics and enhanced mechanical properties, leading to better surface quality. For example, the development of high-strength alloys or composites specifically optimized for AM can help overcome the limitations of currently available materials.

5) **Smart Post-processing Techniques**: With advancements in automation and robotics, smart post-processing techniques are being explored. These techniques involve the use of Artificial Intelligence (AI) and machine learning algorithms to automate and optimize the post-processing steps. By analyzing data on surface defects and mechanical properties, these algorithms can determine the most effective post-processing technique for a given AM part, resulting in improved surface quality.

6) **Surface Modification through Coatings**: Coatings can be applied to AM parts to enhance their surface properties. Functional coatings, such as anti-corrosion coatings or wear-resistant coatings, can improve the surface quality and increase the overall performance of the parts. Future developments may focus on tailored coatings specifically designed for AM materials to further improve surface finish and properties.

It is important to note that these futuristic approaches are still in the research and development stage, and their widespread implementation may take time. However, ongoing advancements in materials science, process control, and automation are driving progress in these areas, and they hold the potential to revolutionize the surface quality and mechanical properties of additively manufactured parts in the future.

10.5 SUMMARY

This chapter provides an overview of the future trends and challenges in post-processing super-finishing techniques for metal Additive Manufacturing (AM). As metal AM becomes more prevalent across industries, there is a growing demand for effective post-processing methods to improve the quality and functionality of AM products. The chapter emphasizes the significance of microstructural features and mechanical properties of metal AM products in selecting appropriate super-finishing techniques.

Various super-finishing techniques are discussed, including abrasive machining, sanding, mass finishing, coating, burnishing, bonding, and electrochemical polishing, that can be employed to enhance the surface finish of AM products. The chapter already addresses the challenges faced by manufacturers in achieving the desired surface finish and proposes potential solutions to overcome these obstacles. Furthermore, the chapter highlighted the importance of future research in advancing the effectiveness of super-finishing processes after metal AM. It concludes by suggesting future research directions to explore and improve these processes, ultimately aiming to enhance the overall quality and performance of metal AM products. Overall, this chapter sheds light on the significance of post-processing super-finishing techniques in metal AM, addresses current challenges, and presents potential avenues for future research and development in this field.

REFERENCES

Augustine, R., Kalva, S. N., Ahmad, R., Zahid, A. A., Hasan, S., Nayeem, A., … Hasan, A. (2021). 3D bioprinted cancer models: Revolutionizing personalized cancer therapy. *Translational Oncology*, *14*(4), 101015. 10.1016/j.tranon.2021.101015

Bowman, L., & Meindl, J. D. (1986). The packaging of implantable integrated sensors. *IEEE Transactions on Biomedical Engineering* (2), 248–255. 10.1109/TBME.1986.325807.

Buchanan, C., & Gardner, L. (2019). Metal 3D printing in construction: A review of methods, research, applications, opportunities and challenges. *Engineering Structures*, *180*, 332–348. 10.1016/j.engstruct.2018.11.045

Carmignato, S., De Chiffre, L., Bosse, H., Leach, R. K., Balsamo, A., & Estler, W. T. (2020). Dimensional artefacts to achieve metrological traceability in advanced manufacturing. *CIRP Annals*, *69*(2), 693–716. 10.1016/j.cirp.2020.05.009

Chohan, J. S., & Singh, R. (2017). Pre and post processing techniques to improve surface characteristics of FDM parts: A state of art review and future applications. *Rapid Prototyping Journal*, *23*(3), 495–513. 10.1108/RPJ-05-2015-0059

Deja, M., Zieliński, D., Kadir, A. Z. A., & Humaira, S. N. (2021). Applications of additively manufactured tools in abrasive machining—A literature review. *Materials*, *14*(5), 1318. 10.3390/ma14051318

Dizon, J. R. C., Gache, C. C. L., Cascolan, H. M. S., Cancino, L. T., & Advincula, R. C. (2021). Post-processing of 3D-printed polymers. *Technologies*, *9*(3), 61. 10.3390/technologies9030061

Duan, Z., Li, C., Ding, W., Zhang, Y., Yang, M., Gao, T., & Ali, H. M. (2021). Milling force model for aviation aluminum alloy: Academic insight and perspective analysis. *Chinese journal of Mechanical Engineering*, *34*(1), 1–35. 10.1186/s10033-021-00536-9

Gray, J., & Luan, B. (2002). Protective coatings on magnesium and its alloys—A critical review. *Journal of Alloys and Compounds*, *336*(1–2), 88–113. 10.1016/S0925-8388(01)01899-0

Horn, T. J., & Harrysson, O. L. (2012). Overview of current additive manufacturing technologies and selected applications. *Science Progress*, *95*(3), 255–282. 10.3184/003685012X13420984463047

Jiang, J., Xu, X., & Stringer, J. (2018). Support structures for additive manufacturing: A review. *Journal of Manufacturing and Materials Processing*, *2*(4), 64. 10.3390/jmmp2040064

Jones, A. R., & Hull, J. B. (1998). Ultrasonic flow polishing. *Ultrasonics*, *36*(1–5), 97–101. 10.1016/S0041-624X(97)00147-9

Kalman, H., Tripathi, N. M., Gabrieli, O. G., & Portnikov, D. (2017). Phase diagrams for pneumatic and hydraulic conveying. In *18th International Conferences on Transport and Sedimentation of Solid Particles, T and S 2017* (pp. 145–152). Wydawnictwo Uniwersytetu Przyrodniczego we Wroclawiu.

Khorasani, M., Ghasemi, A., Rolfe, B., & Gibson, I. (2022). Additive manufacturing a powerful tool for the aerospace industry. *Rapid Prototyping Journal*, *28*(1), 87–100. 10.1108/RPJ-01-2021-0009

Kim, U. S., & Park, J. W. (2019). High-quality surface finishing of industrial three-dimensional metal additive manufacturing using electrochemical polishing. *International Journal of Precision Engineering and Manufacturing-Green Technology*, *6*, 11–21. 10.1007/s40684-019-00019-2

Krishnan, A., & Fang, F. (2019). Review on mechanism and process of surface polishing using lasers. *Frontiers of Mechanical Engineering*, *14*, 299–319. 10.1007/s11465-019-0535-0

Lumay, G., Tripathi, N. M., & Francqui, F. (2019). How to gain a full understanding of powder flow properties, and the benefits of doing so. *ONdrugDelivery*, *2019*, 42–47.

Ma, C. P., Guan, Y. C., & Zhou, W. (2017). Laser polishing of additive manufactured Ti alloys. *Optics and Lasers in Engineering*, *93*, 171–177. 10.1016/j.optlaseng.2017.02.005

Mathew, A., Kishore, S. R., Tomy, A. T., Sugavaneswaran, M., Scholz, S. G., Elkaseer, A., & John Rajan, A. (2023). Vapour polishing of fused deposition modelling (FDM) parts: A critical review of different techniques, and subsequent surface finish and mechanical properties of the post-processed 3D-printed parts. *Progress in Additive Manufacturing*, 1–18. 10.1007/s40964-022-00391-7

Mohanavel, V., Ali, K. A., Ranganathan, K., Jeffrey, J. A., Ravikumar, M. M., & Rajkumar, S. (2021). The roles and applications of additive manufacturing in the aerospace and automobile sector. *Materials Today: Proceedings*, *47*, 405–409. 10.1016/j.matpr.2021.04.596

Mohsen A. (2017). The rise of 3-D printing: The advantages of additive manufacturing over traditional manufacturing, *Business Horizons*, *60*, 677–688. 10.1016/j.bushor.2017.05.011

M Tripathi, N., & S Mallick, S. (2017). Pneumatic conveying of Fly Ash: Bend Models investigation. *Advanced Materials Proceedings*, *2*(8), 526–531.

Neveu, A., Lumay, G., Pillitteri, S., Monsuur, F., Pauly, T., Ribeyre, Q., Francqui, F., Vandewalle, N., & Tripathi, N. M. (2020b). Physical characterization of blends containing mesoporous particles with a focus on electrostatic properties. In *2020 AIChE Spring Meeting and 16th Global Congress on Process Safety*. AIChE.

Neveu, A., Tripathi, N. M., Rigo, O., Francqui, F., & Lumay, G. (2020a). Experimental investigation of spreadability of metal powders in recoating process. *In Proceedings - Euro PM2020 Congress and Exhibition (Proceedings - Euro PM2020 Congress and Exhibition)*. European Powder Metallurgy Association (EPMA).

Pathak, S., & Saha, G. C. (2017). Development of sustainable cold spray coatings and 3D additive manufacturing components for repair/manufacturing applications: A critical review. *Coatings*, *7*(8), 122. 10.3390/coatings7080122

Peng, X., Kong, L., Fuh, J. Y. H., & Wang, H. (2021). A review of post-processing technologies in additive manufacturing. *Journal of Manufacturing and Materials Processing*, *5*(2), 38. 10.3390/jmmp5020038

Ratsimba, A., Zerrouki, A., Tessier-Doyen, N., Nait-Ali, B., André, D., Duport, P., Neveu, A., Tripathi, N., Francqui, F., & Delaizir, G. (2021). Densification behaviour and three-dimensional printing of Y2O3 ceramic powder by selective laser sintering. *Ceramics International*, *47*(6), 7465–7474. https://doi.org/10.1016/j.ceramint.2020.11.087

Salmi, M., Huuki, J., & Ituarte, I. F. (2017). The ultrasonic burnishing of cobalt-chrome and stainless steel surface made by additive manufacturing. *Progress in Additive Manufacturing*, *2*, 31–41. 10.1007/s40964-017-0017-z

Salmi, M. (2021). Additive manufacturing processes in medical applications. *Materials*, *14*(1), 191. 10.3390/ma14010191

Seifi, M., Gorelik, M., Waller, J., Hrabe, N., Shamsaei, N., Daniewicz, S., & Lewandowski, J. J. (2017). Progress towards metal additive manufacturing standardization to support qualification and certification. *Jom*, *69*, 439–455. 10.1007/s11837-017-2265-2

Sharma, A., Babbar, A., Tian, Y., Pathri, B. P., Gupta, M., & Singh, R. (2022). Machining of ceramic materials: A state of the art review. *International Journal on Interactive Design and Manufacturing (IJIDeM)*, 1–21. https://doi.org/10.1007/s12008-022-01016-7

Sharma, A., & Jain, V. (2020). Experimental investigation of cutting temperature during drilling of float glass specimen. In *IOP Conference Series: Materials Science and Engineering* (Vol. 715, No. 1, p. 012050). IOP Publishing.

Sharma, A., Jain, V., & Gupta, D. (2018). Characterization of chipping and tool wear during drilling of float glass using rotary ultrasonic machining. *Measurement*, *128*. 254–263.

Sharma, A., Jain, V., & Gupta, D. (2019a). Tool wear analysis while creating blind holes on float glass using conventional drilling: A multi-shaped tools study. In *Advances in Manufacturing Processes: Select Proceedings of ICEMMM 2018* (pp. 175–183). Springer Singapore.

Sharma, A., Jain, V., & Gupta, D. (2019b). Comparative analysis of chipping mechanics of float glass during rotary ultrasonic drilling and conventional drilling: For multi-shaped tools. *Machining Science and Technology*, *23*(4), 547–568.

Sharma, A., Jain, V., & Gupta, D. (2019c). Multi-shaped tool wear study during rotary ultrasonic drilling and conventional drilling for amorphous solid. *Proceedings of the Institution of Mechanical Engineers, Part E: Journal of Process Mechanical Engineering*, *233*(3), 551–560.

Sharma, A., Jain, V., & Gupta, D. (2021a). Effect of pre and post tempering on hole quality of float glass specimen: For rotary ultrasonic and conventional drilling. *Silicon*, *13*, 2029–2039.

Sharma, A., Jain, V., & Gupta, D. (2021b). Mathematical approach on chipping volume estimation generated during rotary ultrasonic drilling for float glass. *Proceedings of the National Academy of Sciences, India Section A: Physical Sciences*, *92*, 285–291.

Sharma, A., Jain, V., Gupta, D., & Babbar, A. (2020). A review study on miniaturization. In *Advanced Manufacturing and Processing Technology* (First edition, pp. 111–131). CRC Press.

Sharma, A., Kalsia, M., Uppal, A. S., Babbar, A., & Dhawan, V. (2022). Machining of hard and brittle materials: A comprehensive review. *Materials Today: Proceedings*, *50*, 1048–1052.

Tripathi, N. M., Francqui, F., & Lumay, G. (2020). Influence of relative air humidity on the flow property of fine powders. In *Third International Conference on Powder, Granule and Bulk Solids: Innovations and Applications PGBSIA 2020 February 26–28, 2020* (p. 63).

Tripathi, N. M., Francqui, F., Pirenne, T., & Lumay, G. (2019). Measuring food powders electrical properties as a result of anti-static content. In *2019 AIChE Annual Meeting*. American Institute of Chemical Engineers.

Tripathi, N. M., Levy, A., & Kalman, H. (2016). Initial acceleration pressure drop in dilute phase pneumatic conveying system. In *Powder, Granule and Bulk Solids: Innovations and Applications Conference*.

Tripathi, N. M., & Mallick, S. S. (2014). *An Investigation into Pressure Drop Across Bends for Fluidised Densephase Pneumatic Conveying Systems* (Doctoral dissertation).

11 Additive Manufacturing in the Age of Industry 4.0 and Beyond

Mohd Rizwan Jafar, Naveen Mani Tripathi,
Manvendra Yadav, and Daniel Schiochet Nasato

11.1 INTRODUCTION TO ADDITIVE MANUFACTURING AND INDUSTRY 4.0

In the crucible of innovation where digital landscapes meld with tangible production, the manufacturing sphere teeters on the precipice of a groundbreaking epoch. This era heralds not just the amalgamation of technologies but a paradigmatic shift driven by the symbiotic relationship between Additive Manufacturing (AM) and Industry 4.0, reshaping our approach to creation, design, production, and distribution at a foundational level.

11.1.1 THE EVOLUTION OF MANUFACTURING

The tapestry of society has been intricately woven by successive industrial revolutions, from the steam-driven looms that signalled the inaugural transformation to the electrified assembly lines heralding mass production, and onto the computerized automation of the third revolution. Our current vista within the fourth industrial revolution is distinguished by an intricate tapestry of interconnected, intelligent, and digital tools set to redefine the landscape of manufacturing. Industry 4.0 is a quantum leap into a realm where manufacturing techniques coalesce with integrated systems and data analytics to forge environments that are not only smarter and more efficient but also supremely adaptable (Rüßmann et al., 2015). The evolution and trends behind recent machining and additive manufacturing developments have reported great influence on Industry 4.0. (Sharma et al., 2022; Sharma & Jain, 2020; Sharma et al., 2018; Sharma et al., 2019a; Sharma et al., 2019b; Sharma et al., 2019c; Sharma et al., 2021a; Sharma et al., 2021b; Sharma et al., 2020; Sharma et al., 2022; Kalman et al., 2017; Lumay et al., 2019; Tripathi & Mallick, 2017; Neveu et al., 2020a; Neveu et al., 2020b; Ratsimba et al., 2021; Tripathi et al., 2020; Tripathi et al., 2019; Tripathi et al., 2016; Tripathi & Mallick, 2014)

11.1.2 DEFINING ADDITIVE MANUFACTURING

Additive Manufacturing is the process of creating objects in a stratified fashion, layer upon layer from a digital blueprint, utilizing various materials ranging from

DOI: 10.1201/9781003428862-11

metals to polymers. In stark contrast to the subtractive methods of traditional manufacturing, which whittle away material to sculpt a final form, AM adopts an additive approach, meticulously depositing material only where necessary, thus curbing waste and enabling intricate designs (Wohlers & Caffrey, 2015).

Tracing its origins to the 1980s with the emergence of Stereolithography, AM has burgeoned into a plethora of processes like Fused Deposition Modeling, Selective Laser Sintering, and Direct Metal Laser Sintering. The maturation of these processes has expanded AM's utility from rapid prototyping to the actualization of end-use products and specialized tooling.

11.1.3 Overview of Industry 4.0

Industry 4.0 is the banner under which new technologies merge the physical, digital, and biological spheres, affecting all disciplines, economies, and industries. At its core, it features innovations like the Internet of Things, Cyber-Physical Systems, Artificial Intelligence, and Cloud Computing. These technologies empower a real-time connectedness, facilitating a level of flexibility, resource efficiency, and intelligent design previously unseen, thereby crafting a production environment that is at once self-optimizing and futuristic in its adaptability (Kagermann, Wahlster, & Helbig, 2013).

11.1.4 Significance of the Integration

The fusion of AM with Industry 4.0 is profound, representing the quintessence of this revolution's ethos: digitalization, customization, and agility. AM embodies the digital-to-physical conversion, augmenting the design freedom and tailoring of products. Moreover, it is a natural fit within Industry 4.0's digital ecosystem, operating from a digital model and seamlessly integrating into the manufacturing process's digital thread.

Strategically, AM enhances Industry 4.0 by supporting on-demand production, streamlining supply chains, and championing sustainable manufacturing through increased material and energy efficiencies. The alliance of AM and Industry 4.0 is therefore not merely a technological advancement but a radical reimagination of manufacturing philosophy that stands at the vanguard of global innovation, efficiency, and sustainability in production (Ford & Despeisse, 2016).

11.2 INTRODUCTION TO ADDITIVE MANUFACTURING AND INDUSTRY 4.0

11.2.1 Core Technologies of AM

The bedrock of Additive Manufacturing is its diverse and innovative core technologies, each defined by its unique approach to object layer construction.

Stereolithography stands as the pioneer, where the utilization of UV lasers to solidify resin by Charles Hull in 1986 laid the groundwork for precision in AM.

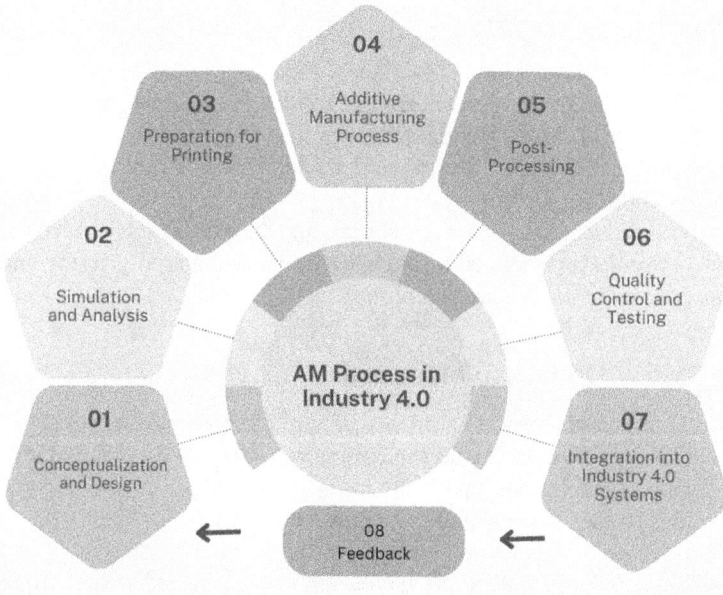

FIGURE 11.1 Flowchart of AM process in Industry 4.0 – This diagram illustrates the comprehensive process flow of Additive Manufacturing within an Industry 4.0 framework, highlighting the stages from design and digital twin simulation to final integration and feedback in smart manufacturing environments.

Selective Laser Sintering and Fused Deposition Modelling followed, introduced by Carl Deckard and Joe Beaman, and Scott Crump, respectively, each contributing to the robustness and complexity possible in AM components (Deckard & Beaman, 1989; Crump, 1992).

The early 2000s saw the maturation of powder bed fusion techniques like Direct Metal Laser Sintering and Electron Beam Melting, known for their capacity to fabricate components with mechanical properties on par with conventional manufacturing, proving indispensable in sectors like aerospace and medicine (Frazier, 2014).

TABLE 11.1
Total Revenue and Market Share

Year	Number of AM Companies	Total Revenue ($B)	Market Share (%)
2018	6500	9.3	1.2
2019	7000	10.4	1.3
2020	7400	12.0	1.5
2021	7800	13.8	1.7
2022	8300	15.6	1.9

11.2.2 Materials in Additive Manufacturing

Material selection remains a cornerstone of AM, with an expanding repertoire that enables boundless creativity:

Metals have surged in AM applications, with titanium alloys and stainless steel being especially prized for their unique properties (Herzog et al., 2016). Polymers continue to be versatile staples due to their printing ease and applicability in various domains like consumer electronics.

Recently, the advent of composite materials and the exploration into bio-inks have opened new frontiers, portended a future where multi-material printing and living tissue fabrication are not outliers but norms.

11.2.3 Digital Design and Modelling

The digital genesis of every AM process lies in Computer-Aided Design (CAD), which enables the intricate digital modelling necessary to exploit AM's potential fully. Software advancements have led to the utilization of parametric and generative design, which leverages algorithms to optimize material placement and structure based on predetermined constraints and performance goals.

Simulation software is becoming increasingly sophisticated, enabling predictive modelling of the AM process, thus ensuring the reliability and quality of printed components (Gibson et al., 2010).

11.3 PILLARS OF INDUSTRY 4.0

11.3.1 The Integration of AM in Industry 4.0 Systems

The amalgamation of AM within Industry 4.0 ecosystems is a natural progression. Smart factories, epitomized by connectedness and data exchange, are enriched by AM's flexibility. AM enhances the cyber-physical systems at the heart of Industry 4.0, granting the capability to respond to feedback and adapt production in real-time (Zhong et al., 2017).

The Digital Thread concept, which weaves a digital continuum throughout the product lifecycle, is bolstered by AM's digital nature. From the initial design to the final build, every modification and process parameter is captured and fed back into the system, enhancing the intelligent learning of these cyber-physical systems.

TABLE 11.2
Evolution of Additive Manufacturing

Year	Key Technological Advances	Number of Patents Registered
2010	Multi-material Printing	300
2012	Advanced Metal AM	450
2014	High-speed Sintering	600
2016	Integrated AM Systems	750
2018	AI in AM	900

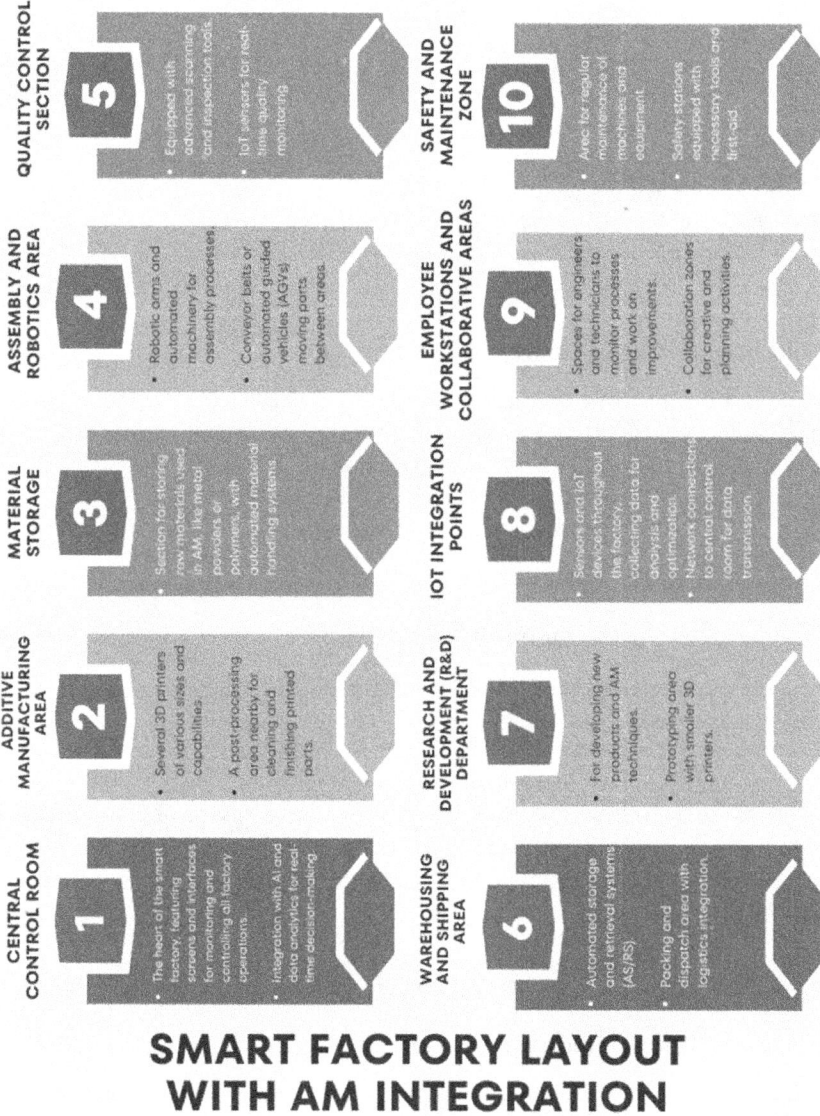

SMART FACTORY LAYOUT WITH AM INTEGRATION

CENTRAL CONTROL ROOM

1

- The heart of the smart factory, featuring screens and interfaces for monitoring and controlling all factory operations.
- Integration with AI and data analytics for real-time decision-making.

ADDITIVE MANUFACTURING AREA

2

- Several 3D printers of various sizes and capabilities.
- A post-processing area nearby for cleaning and finishing printed parts.

MATERIAL STORAGE

3

- Section for storing raw materials used in AM, like metal powders or polymers, with automated material handling systems.

ASSEMBLY AND ROBOTICS AREA

4

- Robotic arms and automated machinery for assembly processes.
- Conveyor belts or automated guided vehicles (AGVs) moving parts between areas.

QUALITY CONTROL SECTION

5

- Equipped with advanced scanning and inspection tools.
- IoT sensors for real-time quality monitoring.

WAREHOUSING AND SHIPPING AREA

6

- Automated storage and retrieval systems (AS/RS).
- Packing and dispatch area with logistics integration.

RESEARCH AND DEVELOPMENT (R&D) DEPARTMENT

7

- For developing new products and AM techniques.
- Prototyping area with smaller 3D printers.

IOT INTEGRATION POINTS

8

- Sensors and IoT devices throughout the factory, collecting data for analysis and optimization.
- Network connections to central control room for data transmission.

EMPLOYEE WORKSTATIONS AND COLLABORATIVE AREAS

9

- Spaces for engineers and technicians to monitor processes and work on improvement.
- Collaboration zones for creative and planning activities.

SAFETY AND MAINTENANCE ZONE

10

- Area for regular maintenance of machines and equipment.
- Safety stations equipped with necessary tools and first-aid.

FIGURE 11.2 Smart Factory Layout with AM Integration – This schematic represents the integration of Additive Manufacturing (AM) within a Smart Factory setup, showcasing the strategic placement of AM units, material storage, quality control, and central monitoring for optimized workflow and production efficiency.

TABLE 11.3
Additive Manufacturing Process

Process Type	Materials Used	Applications	Global Usage (%)
Stereolithography (SLA)	Resin	Prototyping	18
Fused Deposition Modeling (FDM)	Thermoplastic	Tooling	25
Selective Laser Sintering (SLS)	Nylon	Functional Parts	20
Direct Metal Laser Sintering (DMLS)	Metal Alloys	Aerospace	15
Electron Beam Melting (EBM)	Titanium	Medical Implants	8

11.3.2 ROLE OF AM IN CUSTOMIZATION AND FLEXIBILITY

Customization is one of AM's most lauded attributes. It enables the manufacturing of complex, tailored products at no additional cost in tooling or setup, a boon in markets where individualization is increasingly the norm. This flexibility dovetails perfectly with Industry 4.0's imperative for adaptability and quick turnaround, epitomizing the "lot size one" philosophy without compromising efficiency (Thompson et al., 2016).

11.3.3 ADVANCING SUSTAINABILITY THROUGH AM

Environmental considerations are paramount in today's industrial climate. AM rises to this challenge, promoting sustainability through material thriftiness and the optimization of design for minimal waste. Moreover, AM's conducive nature to localized production mitigates the carbon footprint associated with logistics.

In the symbiosis with Industry 4.0, AM's contribution to sustainability extends to energy management within smart factories, where production can be modulated based on renewable energy availability, further entrenching the eco-conscious ethos of this technological union.

11.4 SYNERGY BETWEEN AM AND INDUSTRY 4.0 TECHNOLOGIES

11.4.1 DIGITAL TWIN TECHNOLOGY AND AM

Digital twin technology has become a pivotal aspect of Industry 4.0, offering a real-time digital counterpart of a physical object or process. In Additive Manufacturing (AM), digital twins serve to improve the product lifecycle by allowing for simulation, analysis, and control of the AM process. By simulating the AM process in a virtual environment, engineers can anticipate problems and make adjustments before they occur in the physical world, reducing waste and increasing efficiency (Tao et al., 2018).

11.4.2 AM IN SMART FACTORIES

Smart factories represent the zenith of manufacturing intelligence, where every piece of equipment is interconnected. AM is an integral part of this interconnectedness, offering flexible and adaptive manufacturing capabilities that are fully integrated with factory systems. Smart factories leverage AM for rapid prototyping, tooling, and the direct manufacturing of final products, allowing for a responsive production environment that can adapt to changing market demands (Rayna & Striukova, 2016).

11.4.3 ROLE OF AM IN DIGITAL SUPPLY CHAINS

Digital supply chains utilize digital technologies to enhance the efficiency and responsiveness of supply chain operations. AM plays a critical role by enabling on-demand manufacturing, reducing the need for inventory, and streamlining logistics. It allows for the decentralization of production, closer to the point of consumption, which can significantly impact lead times, reduce transportation costs, and improve carbon footprints (Holmström et al., 2010).

11.4.4 CUSTOMIZATION AND PERSONALIZATION THROUGH AM

AM is uniquely positioned to provide mass customization and personalization in manufacturing. With AM, the cost of individual customization is significantly reduced, as it requires no special tooling and there's virtually no additional cost for complexity. This capability synergizes with Industry 4.0's emphasis on meeting individual customer demands, facilitating a shift from mass production to mass customization, which is becoming increasingly prevalent in markets such as medical implants, dental products, and fashion (Petrovic et al., 2011).

11.5 IMPACT ON DESIGN AND PRODUCT DEVELOPMENT

11.5.1 DESIGN FOR ADDITIVE MANUFACTURING (DFAM)

Design for Additive Manufacturing (DfAM) encompasses the methodologies and tools specific to designing components optimized for Additive Manufacturing processes. It takes advantage of AM's unique capabilities, such as complex geometry construction and material waste minimization. DfAM encourages designers to reconsider traditional design constraints and enables the production of designs that were previously impossible or too costly to fabricate. Components can be lighter, stronger, and more efficient by incorporating complex internal structures, such as lattice or honeycomb patterns that can only be made with AM technologies (Thompson et al., 2016).

11.5.2 COMPLEXITY AND CUSTOMIZATION

Additive manufacturing allows for the creation of complex parts without the cost penalty associated with traditional manufacturing methods, where complexity often equates to higher costs. The inherent nature of AM supports customization at scale, catering to industries that benefit from bespoke solutions, such as healthcare for personalized prosthetics or aerospace for complex, lightweight structures. The ability to produce complex, tailored products rapidly is transforming the way companies approach design and manufacturing, offering them the ability to differentiate through customized products (Rosen, 2007).

11.5.3 Prototyping and Product Testing

The rapid prototyping capabilities of AM accelerate the design and testing phases of product development. Engineers can quickly iterate designs, produce prototypes within hours or days, and conduct testing cycles that significantly shorten the product development process. This speed and flexibility contribute to a more experimental approach to design, where physical prototypes can be produced and tested to validate performance and identify areas for improvement before committing to mass production (Hague et al., 2003).

11.5.4 Iterative Design and Agile Manufacturing

Agile manufacturing is characterized by its adaptability and rapid response to changing customer needs and market conditions. AM plays a critical role in this paradigm by enabling an iterative design process. This approach allows for ongoing refinements and optimization of products even after they have been introduced to the market. Such flexibility in design and production underscores the alignment of AM with agile practices, ensuring that products are not only functional and cost-effective but also continuously evolving (Gibson et al., 2010).

11.6 CASE STUDIES: AM IN DIVERSE INDUSTRIES

The transformative impact of Additive Manufacturing (AM) is best illustrated through case studies in diverse industries, showcasing the technology's potential to revolutionize traditional manufacturing processes, customize products, and significantly cut costs and time.

11.6.1 Aerospace and Defence

11.6.1.1 The LEAP Engine by General Electric

General Electric has not only pushed the boundaries of jet propulsion with its LEAP engine but has also pioneered the use of AM in high-stress, safety-critical aviation components. The fuel nozzles in the LEAP engine, created through Direct Metal Laser Sintering (DMLS), are a highlight. This process consolidates 20 different parts into a single unit, reducing the opportunity for failure due to part junctions. It also cuts the manufacturing time and enables complex internal geometries that enhance performance. Over 30,000 of these nozzles are now in service, marking one of the most significant proofs of AM's maturity in the industry (GE Reports, 2017).

11.6.1.2 F-35 Lightning II by Lockheed Martin

Lockheed Martin uses AM to reduce costs and speed up the production of the F-35 Lightning II. AM enables the rapid prototyping and production of many non-critical components, such as air ducts and brackets. The integration of AM into the production line has been strategic; it allows for localized manufacturing at the point of assembly and reduces the dependency on traditional supply chains, which is crucial in the defence sector where security and speed are paramount (Defense News, 2020).

11.6.2 HEALTHCARE AND BIOMEDICAL

11.6.2.1 Patient-Specific Implants by Oxford Performance Materials

Oxford Performance Materials (OPM) has become a frontrunner in patient-specific implants. Using Polyetherketoneketone (PEKK) and their proprietary OsteoFab technology, OPM can create implants tailored to the patient's anatomy. This customization results in less operative time, better biocompatibility, and potentially better long-term outcomes. A significant case involved creating a device to replace 75% of a patient's skull, showcasing AM's potential to address complex medical challenges (OPM Press Release, 2013).

11.6.2.2 Invisalign by Align Technology

Align Technology has transformed orthodontic care with its Invisalign system. Unlike traditional braces, each Invisalign aligner is unique, designed to fit the patient's mouth perfectly and adjusted throughout the treatment. This is only possible at scale through AM, specifically through Stereolithography (SLA), which enables the mass customization of medical devices. The digital workflow from the patient's mouth to the final product represents the pinnacle of digital dentistry (Align Technology Annual Report, 2020).

11.6.3 AUTOMOTIVE INDUSTRY

11.6.3.1 BMW's Application of AM

BMW has been a leader in adopting AM, not just for prototypes but for production parts. For the i8 Roadster, parts such as the soft-top cover would traditionally be made using injection moulds, which are time-consuming and expensive to produce. By using AM, BMW can produce these parts directly from digital designs, greatly reducing the production lead time and allowing for complex designs that are lighter and use less material (BMW Group Press Release, 2018).

11.6.3.2 Strati by Local Motors

Local Motors introduced Strati, a car with the body and chassis printed in just 44 hours, compared to weeks or months for traditional vehicles. This innovation is not just about the novelty of 3D printing a car; it demonstrates how AM can reduce the part count dramatically, making vehicles simpler to assemble and lighter, which can lead to better fuel efficiency and reduced material waste (Local Motors, 2014).

11.6.4 Consumer Goods

11.6.4.1 Adidas and the Futurecraft 4D

Adidas, in partnership with Carbon, uses Digital Light Synthesis technology to produce the Futurecraft 4D shoe's midsole. This collaboration has resulted in a lattice structure that can be tuned to specific patterns of movement, weight, and foot shape. The ability to adjust cushioning and stability for individual needs presents a new era in athletic performance. The Futurecraft initiative represents a shift in manufacturing towards a more sustainable and personalized approach (Adidas News Stream, 2017).

11.6.5 Architecture and Construction

11.6.5.1 3D Printed Houses by ICON

ICON has leveraged large-scale AM to address housing crises around the world. Their Vulcan printer can produce a house's basic structure in under 24 hours, at a fraction of the cost of traditional construction methods. This rapid construction capability has significant implications for disaster relief, affordable housing, and even extraterrestrial construction scenarios, such as habitats on Mars (ICON, 2020).

11.6.5.2 The MX3D Bridge in Amsterdam

MX3D achieved a milestone in construction and architecture by printing a stainless-steel pedestrian bridge installed in Amsterdam. This case study not only highlights the potential for creating functional, durable structures but also the capability of AM to fabricate intricate designs that are both aesthetically pleasing and structurally sound. The bridge is equipped with sensors to monitor its performance and health, integrating into the Internet of Things (IoT) and providing valuable data for future AM construction projects (MX3D, 2021).

Each of these case studies provides a glimpse into the future possibilities of AM, revealing a pattern of innovation, efficiency, and personalization that could define the next era of manufacturing across industries.

11.7 OPERATIONAL EXCELLENCE WITH AM IN INDUSTRY 4.0

11.7.1 Streamlining Production with AM

Additive Manufacturing (AM) streamlines production by allowing for the direct construction of complex parts from digital files, which eliminates the need for multiple manufacturing processes and assembly operations. By integrating AM into the production workflow, companies can shift from mass production to mass customization, thus meeting diverse customer demands more efficiently.

For instance, Siemens has implemented AM to produce gas turbine blades, a process that has significantly reduced the lead time from design to production, allowing for rapid prototyping and testing. This not only speeds up innovation cycles but also minimizes the energy and material waste associated with traditional manufacturing (Siemens, 2017).

TABLE 11.4
AM in Diverse Industries

Industry	Company/Product	AM Technology Used	Cost Reduction (%)	Production Time Reduction (%)
Aerospace	GE (LEAP Engine)	DMLS	25	30
Defense	Lockheed Martin (F-35)	Various AM Processes	15	20
Healthcare	OPM (Patient-Specific Implants)	PEKK-based AM	20	40
Automotive	BMW (i8 Roadster)	SLS	10	25
Consumer Goods	Adidas (Futurecraft 4D)	Digital Light Synthesis	5	15
Construction	ICON (3D Printed Houses)	Large-scale AM	40	70

11.7.2 INVENTORY AND LOGISTICS MANAGEMENT

The on-demand production capabilities of AM have a profound impact on inventory and logistics management. Instead of keeping large inventories of spare parts, companies can store digital inventories and print parts as needed. This approach reduces warehouse space, decreases capital tied up in inventory, and minimizes waste from unsold products.

A prime example is Airbus, which uses AM to produce parts on-demand for its aircraft, significantly reducing inventory costs and improving service times for aircraft maintenance. The ability to print parts at or near the point of use also greatly simplifies the logistics chain, cutting down on transportation costs and emissions (Airbus, 2020).

11.7.3 MAINTENANCE, REPAIR, AND OVERHAUL (MRO)

In MRO, AM allows for the extension of the lifecycle of products. By producing parts that are no longer available from original manufacturers or by improving the design of existing components, AM ensures the longevity of machinery and equipment. Additionally, AM can facilitate the repair of expensive components, such as those found in aviation and heavy machinery, rather than their replacement.

GE Aviation has leveraged AM for the repair of engine components, which has not only saved costs but also reduced turnaround times. Repaired parts meet or exceed the performance of original components, which is critical for the aviation industry's stringent safety standards (GE Aviation, 2019).

11.7.4 QUALITY CONTROL AND IN-PROCESS VERIFICATION

Quality control is essential in manufacturing, and AM technologies are well-suited to in-process verification due to their layer-by-layer construction method. This allows for real-time monitoring and quality assurance at each stage of the production process. Integrating sensors and feedback mechanisms enables the detection of anomalies early in the production process, thereby reducing waste and ensuring that the final product meets design specifications.

Renishaw has been at the forefront of integrating in-process verification in AM. Their systems incorporate laser scanning to verify the geometry of each layer as it is printed, ensuring that the part is being built correctly and adjusting the process parameters in real-time to correct any deviations (Renishaw, 2018).

11.8 ECONOMIC AND ENVIRONMENTAL CONSIDERATIONS

11.8.1 Cost-Effectiveness of AM

Additive Manufacturing (AM) has shown cost-effectiveness in various sectors, especially where customization and complexity are key factors. For example, in the aerospace sector, AM can lead to a reduction of costs by minimizing material waste and decreasing the weight of components, which translates to fuel savings (Thompson et al., 2016). Moreover, AM's ability to consolidate multiple components into a single part reduces assembly costs (Weller et al., 2015).

11.8.2 Energy Consumption and Sustainability

The sustainability of AM is influenced by the energy consumption of the process itself and the lifecycle impact of manufactured products. While the direct energy use in AM can be high, the overall environmental footprint can be lower when accounting for the entire lifecycle of a product, including the use phase where lightweight structures can result in energy savings (Ford & Despeisse, 2016).

11.8.3 Waste Reduction and Material Efficiency

Waste reduction is one of the key environmental benefits of AM. By building objects layer by layer, AM produces significantly less waste compared to subtractive manufacturing processes. This can be particularly impactful when using expensive or rare materials (Gebler et al., 2014). Material efficiency not only has cost implications but also environmental ones, as it contributes to the conservation of resources (Huang et al., 2016).

11.8.4 The Circular Economy and AM

AM's alignment with the circular economy is evident through its promotion of sustainable design and manufacturing principles. By enabling the production of parts on-demand AM reduces the need for inventory and overproduction, contributing to a reduction in resource consumption and waste. Furthermore, AM technologies can facilitate the repair, refurbishment, and recycling of parts, extending their usable life and supporting the principles of a circular economy.

11.9   FUTURE DIRECTIONS AND CHALLENGES

11.9.1   Technological Advancements on the Horizon

With AM at the cusp of a new era in manufacturing, advancements in printing technology are set to broaden the range of applications. Innovations such as increased print speed, multi-material printing, and improved precision are anticipated. Researchers are also exploring the potential of nanotechnology in AM, which could revolutionize the production of complex structures and materials at a microscopic scale, promising breakthroughs in fields such as electronics and biomedicine (Varotsis, 2023).

11.9.2   Scaling Additive Manufacturing for Mass Production

Transitioning from prototyping to full-scale production is a key challenge for AM. The development of high-throughput AM systems is vital to achieve economies of scale. Continuous Liquid Interface Production (CLIP) and other rapid photopolymerization processes are poised to increase the speed of production without sacrificing resolution, thereby making AM more competitive with traditional manufacturing for mass production (D'Aveni, 2015).

11.9.3   Regulatory and Standardization Challenges

The fragmented landscape of standards and certifications poses significant barriers to the adoption of AM. International standardization efforts are being strengthened to cover more materials, processes, and sectors. Regulatory bodies are called upon to keep pace with the innovation in AM to ensure safety without stifling creativity and progress. A proactive approach to regulation could see the establishment of guidelines that facilitate experimentation while maintaining rigorous standards (ISO/ASTM, 2023).

11.9.4   Workforce Development and Skill Gaps

As AM technologies evolve, so does the need for a skilled workforce. The industry faces a shortage of personnel trained in AM-specific skills, such as design for additive manufacturing (DfAM), process optimization, and materials science. Initiatives like the America Makes programme in the United States aim to address these gaps through partnerships between academia, industry, and government, funding research, and education programmes designed to foster talent in AM (America Makes, 2023).

11.10   BEYOND INDUSTRY 4.0

Looking ahead, additive manufacturing is poised to continue its evolution beyond Industry 4.0. As we enter an era where technology convergence, artificial intelligence, and quantum computing become increasingly prevalent, Additive Manufacturing will likely play a central role in shaping the future of manufacturing.

11.10.1 Bioprinting and Healthcare

The healthcare industry is exploring bioprinting, a specialized form of Additive Manufacturing that creates living tissue and organs. This revolutionary technology has the potential to transform organ transplantation and regenerative medicine.

11.10.2 Space Exploration and Colonization

In the quest for space exploration and colonization, Additive Manufacturing is invaluable. 3D printing can be used to fabricate tools, habitats, and even spare parts on other planets, reducing the need to transport everything from Earth.

11.10.3 Nanoscale Additive Manufacturing

Emerging technologies in nanoscale Additive Manufacturing hold promise for creating intricate and miniature structures with applications in electronics, optics, and medicine. These developments could revolutionize industries where precision at the nanoscale is critical.

11.10.4 Regulatory Frameworks and Ethics

As additive manufacturing advances, there will be a need for robust regulatory frameworks to address safety, quality, and ethical concerns. This includes ensuring the safety of bio-printed organs, establishing guidelines for the use of AI in design optimization, and addressing intellectual property issues in a digital manufacturing landscape.

11.11 CONCLUSION

11.11.1 Summary of AM's Role in Advancing Industry 4.0

Additive Manufacturing (AM) has emerged as a cornerstone of the Fourth Industrial Revolution, often referred to as Industry 4.0. Throughout this chapter, we have explored how AM acts as both a catalyst and a beneficiary of the digital transformation sweeping through the manufacturing sector. It fosters innovation, agility, and customization, all of which are quintessential elements of Industry 4.0.

The integration of AM with other Industry 4.0 technologies, such as the Internet of Things (IoT), big data analytics, and advanced robotics, has initiated a paradigm shift in how products are designed, produced, and distributed. AM's capacity for complexity and customization has unlocked new product designs and business models, presenting a clear divergence from the constraints of traditional manufacturing.

In reviewing the synergies between AM and Industry 4.0, we have noted how these technologies combine to create smart factories where digital cohesion and physical production converge. AM's flexibility is a natural fit for the digital thread that runs through Industry 4.0, connecting disparate processes and enabling a seamless flow of data across the product lifecycle.

11.11.2 Final Thoughts on the Future of Digital Manufacturing

Looking ahead, the trajectory of digital manufacturing is poised to ascend further with AM at its core. The prospective technological advancements in AM promise to enhance materials, processes, and the scope of applications. However, the full potential of AM within Industry 4.0 will only be realized if the accompanying challenges are addressed head-on. This includes scaling production for mass markets, navigating the evolving regulatory landscape, standardizing processes across borders, and filling the growing skills gap in the workforce.

The AM industry must also continue to forge partnerships among stakeholders in business, academia, and government to cultivate an ecosystem that encourages innovation while also establishing robust educational programmes to prepare the next generation of engineers and designers for a future where digital manufacturing is the norm.

In sum, AM is not merely a standalone innovation but a transformative force that propels the entire manufacturing sector towards a more agile, sustainable, and customized future. It is a symbol of the future of digital manufacturing – a future that is being written today.

REFERENCES

Adidas News Stream. (2017). Adidas futurecraft 4D.

Airbus. (2020). Airbus and 3D printing for aircraft parts.

Align Technology Annual Report. (2020). Innovations in orthodontics.

America Makes. (2023). America makes: The National Additive Manufacturing Innovation Institute. Retrieved from America Makes website

Blériot, J., & Johnson, C. (2013). *A New Dynamic Effective Business in a Circular Economy*. Ellen MacArthur Foundation Publishing.

BMW Group Press Release. (2018). Additive Manufacturing at BMW.

Crump, S. S. (1992). *U.S. Patent No. 5,121,329*. U.S. Patent and Trademark Office.

D'Aveni, R. (2015). The 3-D printing revolution. *Harvard Business Review, 93*(5), 40–48.

Deckard, C. R., & Beaman, J. J. (1989). Selective laser sintering with assisted powder handling. U.S. Patent No. 4,863,538.

Defense News. (2020). F-35 and additive manufacturing.

Ford, S., & Despeisse, M. (2016). Additive manufacturing and sustainability: An exploratory study of the advantages and challenges. *Journal of cleaner Production, 137*, 1573–1587. DOI: 10.1016/j.jclepro.2016.04.150

Frazier, W. E. (2014). Metal additive manufacturing: A review. *Journal of Materials Engineering and performance, 23*, 1917–1928. DOI: 10.1007/s11665-014-0958-z

GE Aviation. (2019). The use of additive manufacturing in MRO.

GE Reports. (2017). The LEAP engine.

Gebler, M., Uiterkamp, A. J. S., & Visser, C. (2014). A global sustainability perspective on 3D printing technologies. *Energy Policy, 74*, 158–167. DOI: 10.1016/j.enpol.2014.08.033

Gibson, I., Rosen, D. W., Stucker, B., Gibson, I., Rosen, D. W., & Stucker, B. (2010). Direct digital manufacturing. *Additive Manufacturing Technologies: Rapid Prototyping to Direct Digital Manufacturing, 378–399*. DOI: 10.1007/978-1-4419-1120-9

Gibson, I., Rosen, D. W., & Stucker, B. (2010). *Additive Manufacturing Technologies: 3D Printing, Rapid Prototyping, and Direct Digital Manufacturing*. Springer.

Hague, R., Campbell, I., & Dickens, P. (2003). Implications on design of rapid manufacturing. *Proceedings of the Institution of Mechanical Engineers, Part C: Journal of Mechanical Engineering Science, 217*(1), 25–30. DOI: 10.1243/095440603762554570

Herzog, D., Seyda, V., Wycisk, E., & Emmelmann, C. (2016). Additive manufacturing of metals. *Acta Materialia, 117*, 371–392. DOI: 10.1016/j.actamat.2016.07.019

Holmström, J., Partanen, J., Tuomi, J., & Walter, M. (2010). Rapid manufacturing in the spare parts supply chain: Alternative approaches to capacity deployment. *Journal of Manufacturing Technology Management, 21*(6), 687–697. DOI: 10.1108/JMTM-03-2015-0022

Huang, R., Riddle, M., Graziano, D., Warren, J., Das, S., Nimbalkar, S., … Masanet, E. (2016). Energy and emissions saving potential of additive manufacturing: The case of lightweight aircraft components. *Journal of Cleaner Production, 135*, 1559–1570. DOI: 10.1016/j.jclepro.2013.07.057

ICON. (2020). Building homes using 3D printing.

ISO/ASTM. (2023). ISO/ASTM 52900:2023 additive manufacturing — General principles — Terminology. International Organization for Standardization. Retrieved from ISO/ASTM website.

Kagermann, H., Wahlster, W., & Helbig, J. (2013). Recommendations for implementing the strategic initiative INDUSTRIE 4.0. Final report of the Industrie 4.0 Working Group.

Kalman, H., Tripathi, N. M., Gabrieli, O. G., & Portnikov, D. (2017). Phase diagrams for pneumatic and hydraulic conveying. In *18th International Conferences on Transport and Sedimentation of Solid Particles, T and S 2017* (pp. 145–152). Wydawnictwo Uniwersytetu Przyrodniczego we Wroclawiu.

Local Motors. (2014). The Strati - The world's first 3D printed car.

Lumay, G., Tripathi, N. M., & Francqui, F. (2019). How to gain a full understanding of powder flow properties, and the benefits of doing so. *ONdrugDelivery, 2019*, 42–47.

M Tripathi, N., & S Mallick, S. (2017). Pneumatic conveying of Fly Ash: Bend Models investigation. *Advanced Materials Proceedings, 2*(8), 526–531.

MX3D. (2021). The MX3D bridge.

Neveu, A., Lumay, G., Pillitteri, S., Monsuur, F., Pauly, T., Ribeyre, Q., Francqui, F., Vandewalle, N., & Tripathi, N. M. (2020b). Physical characterization of blends containing mesoporous particles with a focus on electrostatic properties. In *2020 AIChE Spring Meeting and 16th Global Congress on Process Safety*. AIChE.

Neveu, A., Tripathi, N. M., Rigo, O., Francqui, F., & Lumay, G. (2020a). Experimental investigation of spreadability of metal powders in recoating process. In *Proceedings - Euro PM2020 Congress and Exhibition (Proceedings - Euro PM2020 Congress and Exhibition)*. European Powder Metallurgy Association (EPMA).

OPM Press Release. (2013). OPM's patient-specific cranial devices.

Petrovic, V., Vicente Haro Gonzalez, J., Jordá Ferrando, O., Delgado Gordillo, J., Ramón Blasco Puchades, J., & Portolés Griñan, L. (2011). Additive layered manufacturing: Sectors of industrial application shown through case studies. *International Journal of Production Research, 49*(4), 1061–1079. DOI: 10.1080/00207543.2010.499360

Rayna, T., & Striukova, L. (2016). From rapid prototyping to home fabrication: How 3D printing is changing business model innovation. *Technological Forecasting and Social Change, 102*, 214–224. DOI: 10.1016/j.techfore.2015.07.023

Ratsimba, A., Zerrouki, A., Tessier-Doyen, N., Nait-Ali, B., André, D., Duport, P., Neveu, A., Tripathi, N., Francqui, F., & Delaizir, G. (2021). Densification behaviour and three-dimensional printing of Y2O3 ceramic powder by selective laser sintering. *Ceramics International, 47*(6), 7465–7474. https://doi.org/10.1016/j.ceramint.2020.11.087

Renishaw. (2018). In-Process Verification in Additive Manufacturing.

Rosen, D. W. (2007). Design for additive manufacturing: A method to explore unexplored regions of the design space. In *Eighteenth Annual Solid Freeform Fabrication Symposium*. University of Texas at Austin, 402–415.

Rüßmann, M., Lorenz, M., Gerbert, P., Waldner, M., Justus, J., Engel, P., & Harnisch, M. (2015). Industry 4.0: The future of productivity and growth in manufacturing industries. *Boston consulting group*, 9(1), 54–89.

Sharma, A., Babbar, A., Tian, Y., Pathri, B.P., Gupta, M., & Singh, R. (2022). Machining of Ceramic Materials: A state of the art review. *International Journal on Interactive Design and Manufacturing (IJIDeM)*, 1–21. DOI: https://doi.org/10.1007/s12008-022 -01016-7

Sharma, A., & Jain, V. (2020). Experimental investigation of cutting temperature during drilling of float glass specimen. In *IOP Conference Series: Materials Science and Engineering* (Vol. 715, No. 1, p. 012050). IOP Publishing.

Sharma, A., Jain, V., & Gupta, D. (2018). Characterization of chipping and tool wear during drilling of float glass using rotary ultrasonic machining. *Measurement*, 128, 254–263.

Sharma, A., Jain, V., & Gupta, D. (2019a). Tool wear analysis while creating blind holes on float glass using conventional drilling: A multi-shaped tools study. In *Advances in Manufacturing Processes: Select Proceedings of ICEMMM 2018* (pp. 175–183). Springer Singapore.

Sharma, A., Jain, V., & Gupta, D. (2019b). Comparative analysis of chipping mechanics of float glass during rotary ultrasonic drilling and conventional drilling: For multi-shaped tools. *Machining Science and Technology*, 23(4), 547–568.

Sharma, A., Jain, V., & Gupta, D. (2019c). Multi-shaped tool wear study during rotary ultrasonic drilling and conventional drilling for amorphous solid. *Proceedings of the Institution of Mechanical Engineers, Part E: Journal of Process Mechanical Engineering*, 233(3), 551–560.

Sharma, A., Jain, V., & Gupta, D. (2021a). Effect of pre and post tempering on hole quality of float glass specimen: For rotary ultrasonic and conventional drilling. *Silicon*, 13, 2029–2039.

Sharma, A., Jain, V., & Gupta, D. (2021b). Mathematical approach on chipping volume estimation generated during rotary ultrasonic drilling for float glass. *Proceedings of the National Academy of Sciences, India Section A: Physical Sciences*, 92, 285–291.

Sharma, A., Jain, V., Gupta, D., & Babbar, A. (2020). A review study on miniaturization. In *Advanced Manufacturing and Processing Technology* (First edition, pp. 111–131). CRC Press.

Sharma, A., Kalsia, M., Uppal, A. S., Babbar, A., & Dhawan, V. (2022). Machining of hard and brittle materials: A comprehensive review. *Materials Today: Proceedings*, 50, 1048–1052.

Siemens. (2017). Siemens gas turbine blades and additive manufacturing.

Tao, F., Cheng, J., Qi, Q., Zhang, M., Zhang, H., & Sui, F. (2018). Digital twin-driven product design, manufacturing and service with big data. *The International Journal of Advanced Manufacturing Technology*, 94, 3563–3576. DOI: 10.1007/s00170-017-0233-1

Thompson, M. K., Moroni, G., Vaneker, T., Fadel, G., Campbell, R. I., Gibson, I., … Martina, F. (2016). Design for additive manufacturing: Trends, opportunities, considerations, and constraints. *CIRP Annals*, 65(2), 737–760. DOI: 10.1016/j.cirp.2016.05.004

Tripathi, N. M., Francqui, F., & Lumay, G. (2020). Influence of relative air humidity on the flow property of fine powders. In *Third International Conference on Powder, Granule and Bulk Solids: Innovations and Applications PGBSIA 2020 February 26–28, 2020* (p. 63).

Tripathi, N. M., Francqui, F., Pirenne, T., & Lumay, G. (2019). Measuring food powders electrical properties as a result of anti-static content. In *2019 AIChE Annual Meeting*. American Institute of Chemical Engineers.

Tripathi, N. M., Levy, A., & Kalman, H. (2016), Initial acceleration pressure drop in dilute phase pneumatic conveying system. In *Powder, Granule and Bulk Solids: Innovations and Applications Conference*.

Tripathi, N. M., & Mallick, S. S. (2014). *An Investigation into Pressure Drop Across Bends for Fluidised Densephase Pneumatic Conveying Systems* (Doctoral dissertation).

Varotsis, A. (2023). Nanotechnology in 3D printing: The new frontier in additive manufacturing. *Advanced Materials Technologies, 8*(1), 2000456. DOI: 10.1002/admt.202000456

Weller, C., Kleer, R., & Piller, F. T. (2015). Economic implications of 3D printing: Market structure models in light of additive manufacturing revisited. *International Journal of Production Economics, 164*, 43–56. DOI: 10.1016/j.ijpe.2015.02.020

Wohlers, T. T., & Caffrey, T. (2015). *Wohlers Report 2015: 3D Printing and Additive Manufacturing State of the Industry Annual Worldwide Progress Report*. Wohlers Associates.

Zhong, R. Y., Xu, X., Klotz, E., & Newman, S. T. (2017). Intelligent manufacturing in the context of industry 4.0: A review. *Engineering, 3*(5), 616–630. DOI: 10.1016/J.ENG.2017.05.015

Index

For Product Safety Concerns and Information please contact our EU
representative GPSR@taylorandfrancis.com
Taylor & Francis Verlag GmbH, Kaufingerstraße 24, 80331 München, Germany

www.ingramcontent.com/pod-product-compliance
Lightning Source LLC
Chambersburg PA
CBHW060357220326
41598CB00023B/2946

9 781032 550664